The Biggest Transfer of Wealth in the Age of Cryptocurrency

By: Xavier Aponte

Published by Diamond League Publishing Corporation

Apontecryptocompliance@gmail.com

Hardcopy ISBN # 979-8-234-13139-3

Paperback ISBN # 979-8-234-13140-9

https://juicefundamentals.com/

Table of Contents

Disclaimer

The contents of this book are based on my personal memories, experiences, and recollections. To respect the privacy of individuals and to preserve anonymity, I have altered certain names, locations, physical descriptions, occupations, and other identifying details. These changes have been made to ensure that any individuals or places discussed cannot be easily recognized, while still conveying the essence of the experiences described. This book has taken a few years to put together and publish, so a lot of the information contained within this book about the cryptocurrency space may be outdated.

This book is a narrative and does not substitute for any form of professional advice. It is important to understand that the information presented herein is for informational and educational purposes only. It should not be considered a replacement for legal, financial, or other professional advice. Readers are strongly encouraged to consult with appropriate professionals when seeking advice related to their personal circumstances or other specific situations.

While every effort has been made to ensure the accuracy and reliability of the content, the information presented in this book and any associated materials may contain inaccuracies, typographical errors, or outdated information. The author and publisher make no guarantees regarding the accuracy, completeness, or timeliness of the content, and they are not responsible for any errors or omissions that may be present. Readers should be aware that the information is subject to change, and improvements or revisions may be made to the content at any time without notice.

Additionally, any advice or opinions expressed in this book is not intended to influence personal, legal, financial, or business decisions. Any references to financial matters are strictly for educational purposes and should not be construed as financial advice. The author and publisher do not assume any responsibility for any financial or other decisions made based on the content of this book. Readers are recommended to consult a qualified professional, such as a financial advisor, before making any significant financial decisions. Please note that past performance in any area, including finance, does not guarantee future outcomes or results.

The information provided in this book is intended for general informational use only. If you need specific advice related to your personal finances, or legal matters, you should seek out a professional with expertise in the relevant field. Under no circumstances should any aspect of this book be relied upon as a sole resource for making important life decisions. The author and publisher disclaim any liability for damages, losses, or other consequences resulting from the use or misuse of the information contained in this book.

A Note from the Author

Before I dive into my dedications, I want to take a moment to clarify something important. Every word written in this book comes from my firsthand experiences in the cryptocurrency space, both within my business and my personal trading life. I am not a licensed financial advisor. For those of you seeking specific financial advice, I strongly encourage you to consult with a qualified financial advisor. Cryptocurrency is an incredibly volatile market, and it is essential that you do your own research before making any investment decisions. The information shared in this book is for educational purposes only. My hope is that you find value in the insights shared, and that this book guides you towards discovering the "hidden gems" that I refer to as JUICE fundamentals. JUICE stands for *Join Us In Crypto Expansion,* a concept I believe captures the essence of the exciting opportunities in the cryptocurrency space. May this book bring you precious nuggets of gold and help you build confidence as you explore the world of digital currency. I honestly believe that we are currently witnessing what I consider to be the Biggest Transfer of Wealth in modern history, and cryptocurrency is at the heart of this movement.

Dedication to God and Jesus Christ

As this is my second book, it takes a completely different approach compared to my first, which was my autobiography, *Can't Break My Spirit*. In this book, I wanted to impart everything I've learned about the world of cryptocurrency—knowledge that has empowered not just my own journey, but the lives of many others. But before I get into the specifics of this journey, I must take a moment to give thanks to my God and Savior, Jesus Christ. Without His guidance and grace, the very idea of drafting this book or even starting my crypto consulting business would never have come to fruition. All of this began with a simple idea back in 2016. I was lying in bed when the thought of launching a crypto consulting business entered my mind. At that time, I didn't have the expertise I possess today, nor the financial resources to make it happen, but I had something far more powerful: commitment. I made the decision to take massive action and learn everything I could about the cryptocurrency space.

The early days were challenging. I had no clear direction and stepped into the industry simply with belief, ambition, and the willingness to learn. I began by offering consulting services at a low price yet still struggled to attract clients. The setbacks were real, and the lessons were tough. But even through those moments, my passion never faded, and my vision for what was possible never wavered. I stayed committed to growth, sharpening my skills, refining my strategy, and building real value. Today, I command premium fees for exclusive one-on-one consulting, working closely with clients who value expertise, strategy, and true transformation. Vastly different from those first days, and I can only attribute these blessings to God's intervention. The journey has been filled with victories, and every single one of those victories is His. I am deeply grateful for His guidance, and for the strength He has given me to help others in their journeys—just as He has helped me. God, even in moments when I'm not at my best, You've never forsaken me. Thank You for allowing me to touch lives and help others in gaining the knowledge that will help their families, just as You have helped me and mine.

Dedication to My Grandparents, Jose and Debra

My next dedication is to my grandparents, Jose and Debra. I love you both with all my heart. You have been an incredible source of love, guidance, and support throughout my life. Debra, you have been such a massive blessing to not only my life but also my business. From helping me with my blogs on my website to supporting my dreams in countless ways, you both have been a constant encouragement. Your unwavering belief in me has been a driving force behind my success, and your contributions have played a pivotal role in helping me continue to push forward.

Together you have provided me with the stability and wisdom needed to navigate through my many challenges. I am forever thankful to God for allowing me to have you both in my life. Your influence has shaped me into who I am today, and I know it has also positively impacted my brother, Anthony. I love you both dearly, and I cannot express enough how much I appreciate everything you've done for me and for all the ways you continue to bless my life.

Dedication to My Developers and Team

Lastly, I want to dedicate a special thank you to my developers and everyone who has contributed to making this book and my crypto consulting business a reality. Without your expertise, support, and dedication, I would not have been able to create the types of products and services that have positively impacted thousands of people worldwide. Through our work together, we have created life-changing opportunities—helping individuals break free from the traditional 9-5 job grind and showing them the incredible potential of the cryptocurrency market.

The results have been astonishing—creating new millionaires and helping people gain financial freedom that they previously thought was out of reach. More than just financial gains, we have helped people understand the benefits of the cryptocurrency space and how to navigate it for their advantage.

The opportunity to share the advantages of this new digital economy with the world has been one of the most rewarding experiences of my life, and I am incredibly grateful to have each of you by my side in this journey. This book is a testament to our collective efforts, and I hope it serves as a guide for anyone looking to explore the world of crypto and unlock the potential within this transformative industry.

As you read these words, I hope this book serves as not just a collection of knowledge, but as a beacon to guide you through the exciting and ever-evolving world of cryptocurrency.

May it empower you, enrich your life, and inspire you to take bold steps towards financial freedom and success.

-Xavier Aponte

Introduction

Are We Moving Toward the Age of Digital Cash?

As technology continues to advance at an unprecedented pace, we find ourselves on the brink of a monumental shift in the way we think about money and financial systems. We are moving into an era where digital cash is becoming a reality. Picture a world where you can earn cryptocurrency simply by playing games in virtual lands. These aren't just fantasies from science fiction anymore; they are emerging realities that many are already experiencing.

Advancements in quantum computing and the compatibility of artificial intelligence are making what seemed impossible just a few years ago possible today. For those of you reading this book, I am here to share my journey through this revolutionary space. At 28 years old, I reached a significant milestone, earning cryptocurrency through my work with SmartTrader at the peak of Bitcoin's all-time high (ATH) in 2021 when it hit $69,000. My involvement in the cryptocurrency world began in 2016, and since then, I've been building a business that's not only transformed my life but the lives of over 4,000 students around the world.

Out of these students, I'm proud to say that some have become millionaires, and have quit their 9-to5 jobs to pursue a life of freedom by working from home. But before I dive deeper into my journey and the insights I've learned along the way, I want to take a moment to thank you. Thank you for investing in this book and taking the time to read it. My hope is that the information shared here will serve as a valuable resource for you, one you can refer to time and again, as you begin your own journey in the world of cryptocurrency.

A Guide, Not Just a Book

This book is not your average cryptocurrency guide. It's not a simple introduction to blockchain technology, nor is it just another brief explanation of Bitcoin. In fact, many books on cryptocurrency offer surface-level insights that are either overly technical or difficult for newcomers to understand. This book is different. It is based on my firsthand experiences that started in 2016 when I first entered the crypto currency space. Since then, I've learned through trial and error, through successes and failures, and I'm here to share everything I've learned with you.

Cryptocurrency can be overwhelming for beginners. Performing a simple search on Google can often lead to more questions than answers. But in this book, you will find my personal roadmap, a step-by step guide to understanding cryptocurrency, decentralized finance (DeFi), and how to make smart investments that can change your life. My goal is not only to teach you the fundamentals but to help you navigate the biggest transfer of wealth we've ever seen. This is the dawn of a new financial age, and it's happening right now.

Traditional banking systems and centralized financial institutions have long held the power over how we manage money. But blockchain technology and decentralized cryptocurrencies are changing that, offering people the power to control their own financial destiny. This is a massive shift, and countries like El Salvador are leading the way by adopting Bitcoin as legal tender and giving their citizens the freedom to operate outside traditional financial systems. The unbanked now can be their own bank. This is just the beginning, and the potential for growth is staggering.

In the years to come, we will hear more stories of individuals who turned small investments into large fortunes. Some people may even make millions from modest investments of just a few hundred dollars. And believe me, I've seen it happen. I've experienced it myself.

My Journey to Crypto Success

When my own "overnight success" story unfolded, it was more than just a stroke of luck—it was confirmation that the principles of cryptocurrency investing truly work. Before the so-called "overnight success," I had built a successful consulting business. Over time, this business grew into a thriving community, many of whom found immense success in this space.

At the time of writing this, students have become millionaires. These are real people, from all walks of life, took what they learned from my consulting program and used it to generate life-changing wealth. Others have used the knowledge to quit their jobs and become financially independent, stepping away from the grind of the traditional 9-to-5 world.

One of the most rewarding aspects of my journey has been watching my students' successes unfold. I've seen students take just $500 and turn it into $30,000 to $100,000 by investing in promising DeFi projects. This is not hypothetical; it's a real and repeatable process that I've personally guided others through. It's one of the most exciting aspects of the cryptocurrency space—the ability to make significant profits even with relatively small investments.

But it doesn't stop there. Over the years, I've helped create amazing gains in the crypto world. My consulting business has given me the privilege of working with amazing people, many of whom were friends, family, and loved ones, helping them grow their wealth and change their lives. I've worked with clients whose portfolios started at just $10,000 and helped them build substantial wealth, turning their initial investments into fortunes and have created their own stories.

The Future of Cryptocurrency and Your Role in It

The future of cryptocurrency is incredibly bright, and we are only scratching the surface of what is possible. The decentralized nature of blockchain technology is challenging established financial systems and creating an entirely new financial ecosystem. The world is changing, and cryptocurrency is at the forefront of this change. The early adopters who understand this technology and its potential will be the ones to reap the greatest rewards.

Throughout this book, you will learn not only how to make smart investments but also how to position yourself for long-term success in this new digital economy. Cryptocurrency isn't just a way to make money; it's a way to take control of your financial future, break free from traditional systems, and capitalize on the greatest wealth transfer the world has ever seen.

I invite you to think about the possibilities that lie ahead and how cryptocurrency can change your life. By the end of this book, I hope you'll have the knowledge, tools, and confidence to navigate this exciting new world and make decisions that will propel you toward financial freedom.

Conclusion

As you read this book, remember that cryptocurrency is more than just a passing trend. It is a transformative force that is reshaping the global financial landscape. The opportunities are immense, but so are the risks. My mission in writing this book is to provide you with the knowledge, insights, and strategies you need to succeed in this new age of digital finance. It is my hope that by following the principles laid out in these pages, you will not only understand the power of cryptocurrency but also feel empowered to act and seize the opportunities that lie ahead.

Witnessing the rise of cryptocurrency millionaires has been nothing short of incredible. seeing friends and students, many of whom were once in the same position as I was, succeeding in the cryptocurrency space is the ultimate reward The levels of success and the heartwarming testimonials I've received speak volumes about the transformative power of digital currencies, and now, the incredible hype surrounding NFTs and the Metaverse. This book will not just explain how to buy and sell cryptocurrencies but will dive deep into the strategies that I have personally used and taught my students to maximize their gains.

I will share with you the diverse ways to take advantage of cryptocurrency debit and credit cards, and how to leverage platforms to stake your cryptocurrencies and earn a significant Annual Percentage Yield (APY) every year. This book is designed to walk you through my firsthand experiences, showing you how I have successfully navigated the digital currency world and helped others do the same. It will also guide you through setting up secure accounts such as hardware wallets (also known as cold storage) to safely store your assets. Whether you're using centralized exchanges or learning about decentralized exchanges (DEXs), I'll provide practical advice and step-by-step instructions to help you build a solid foundation.

Before we go further, I want to remind you that I am not a financial advisor—the information in this book comes from my firsthand experiences and the success stories of the students I've been fortunate enough to mentor. It is essential that you always do your own research before making any financial decisions.

Thank you for trusting me to be your guide on this journey. The future is digital, and I'm excited for you to be a part of it.

Chapter 1

What is Blockchain and What Influence Does It Have Today

In the world of cryptocurrency, blockchain technology is the foundation that makes it all possible. Understanding the blockchain and how it works is crucial for anyone entering this space, whether you are an investor, trader, or just a curious reader. To fully appreciate the power of digital currencies like Bitcoin, Ethereum, and others, it's important to understand how the blockchain functions as the "Ledger of Trust."

What is Blockchain?

The blockchain concept was introduced by an individual—or group—known as Satoshi Nakamoto, whose identity remains shrouded in mystery. In 2008, Nakamoto released the groundbreaking whitepaper titled "Bitcoin: A Peer-to-Peer Electronic Cash System." This paper came at a time when the global financial system was facing a crisis. Markets were crashing, and people were losing trust in centralized financial institutions. Bitcoin was born out of this turmoil, with a mission to decentralize financial systems and offer a peer-to-peer version of electronic cash free from government control.

But what exactly is blockchain? According to Euromoney.com, a blockchain is "essentially a digital ledger of transactions that are duplicated and distributed across the entire network of computer systems on the blockchain." Each "block" in the blockchain contains a record of transactions. As new transactions occur, they are added to every participant ledger, making it almost impossible for anyone to alter past transactions without being detected. This distributed ledger technology (DLT) makes blockchain incredibly secure, transparent, and decentralized.

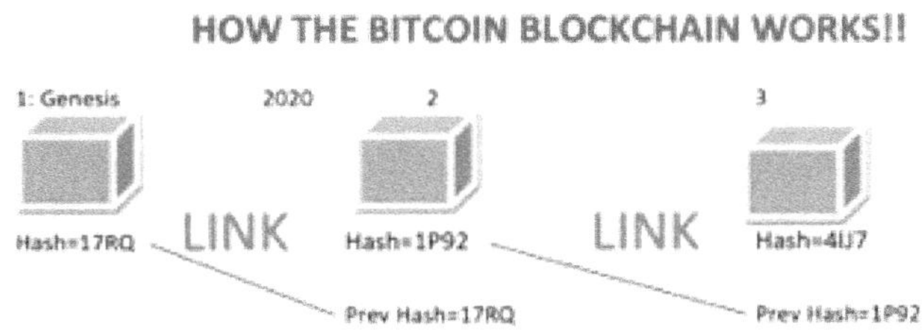

Each Block:

1) Data: Send Address, Receive Address, Time, Amount
2) Hash: Identification of the block
3) Hash Of Previous Block

Every blockchain project starts with what is known as the Genesis Block — the very first block in the chain. After the Genesis Block is created, subsequent blocks are added, and transaction data is sent and received across the decentralized network.

The beauty of blockchain lies in its transparency and security. Every transaction is recorded, and once added to the chain, it becomes part of an immutable ledger. This makes blockchain technology not only trustworthy but also secure, as the data within it cannot be easily altered or erased.

For example, if you want to look up a Bitcoin transaction, you can use the transaction ID (TXID) or the Ethereum address. By searching the blockchain on a platform like Etherscan, you can trace every transaction made with that address as it is publicly visible for anyone to see. This transparency and immutability are some of the key reasons why blockchain is being hailed as a revolution in digital security.

How to Read a Transaction on Etherscan

Here is a breakdown of how one can follow a transaction on Etherscan:

1. Go to Etherscan.io.
2. Enter the transaction hash (TxHash) in the search bar.
3. Review the details explained below.

Transaction Hash (TxHash)

- This is like a receipt number for the transaction. It's unique to each transaction.
- Example:
  0x5d1e5c0a2b6f3c8b3f1d2a4b5e6f7a8b2f3c9d0e1a2b3c4d5e6f7a8b9c0d1e2f

Status

- Indicates whether the transaction was successful, failed, or is still pending. Common statuses:
  - **Success** – The transaction was confirmed.
  - **Failed** – It didn't go through, possibly due to low gas. ○ **Pending** – It's still waiting for confirmation.

Block

- The block number in which the transaction was recorded.
- Think of a block as a "page" in the blockchain ledger.
- Example: Block 19485723

Timestamp

- The exact date and time the transaction was confirmed.
- Example:
  Feb-28-2025 10:15:32 AM UTC

From (Sender Address)

- The Ethereum wallet that initiated the transaction.
- Example:
  0x1234abcd5678efgh9012ijkl3456mnop7890qrst

To (Recipient Address)

- The Ethereum address receiving the funds or interacting with the contract.
- If it's a smart contract, this is the contract's address.
- Example:
  0x5678mnop9012qrst3456abcd7890efgh1234ijkl

Value

(Amount/Sent)

- The amount of ETH transferred in the transaction.
- Example:
  0.5 ETH

Transaction Fee (Gas Fee)

- The fee paid to miners/validators to process the transaction.
- Example:
  0.00021 ETH

Gas Price

- The cost per unit of gas, measured in Gwei (1 Gwei = 0.000000001 ETH).
- Higher gas prices help transactions get confirmed faster.
- Example:
  30 Gwei

Gas Limit & Gas Used

- **Gas Limit:** The maximum gas the sender is willing to pay for execution.
- **Gas Used:** The actual gas consumed by the transaction.
- Example:
  - **Gas Limit:** 21000 (standard for ETH transfers)
  - **Gas Used:** 21000 (actual gas used)

Nonce

- The number of transactions sent from the sender's address.
- Example:
  12 (meaning this was the 12th transaction from this address)

Input Data

- If sending ETH: This field is empty.
- If interacting with a smart contract: This contains encoded data, which is used to trigger functions in the contract.
- Example (empty for ETH transfer):
  0x

Exploring the Future of Financial Transactions: Blockchain and Cryptocurrencies

One of the core problems blockchain seeks to solve is the lack of security in traditional banking systems. Data breaches are becoming all too common, with governments and businesses falling victim to hackers. This ongoing erosion of privacy is one of the driving forces behind the creation of Bitcoin and other cryptocurrencies. By decentralizing currency and removing the need for intermediaries, blockchain gives people control over their money without relying on centralized institutions.

Blockchain is also playing a pivotal role in empowering countries that have been excluded from the traditional banking system due to high inflation or prohibitively expensive banking fees. In nations like El Salvador and Venezuela, cryptocurrency is gaining acceptance as a legitimate form of payment, offering people an alternative to unstable national currencies. This is especially important in countries with rampant inflation or struggling economies, where blockchain technology provides a means for citizens to manage their financial futures without relying on a centralized banking system.

For nations with unstable currencies and banking systems, blockchain provides a solution. In countries where inflation is rampant, traditional banking services are often too costly or even unavailable to the average person. Cryptocurrencies and blockchain enable people to become their own bank and participate in the global economy without the barriers imposed by central authorities. Like El Salvador, many countries are embracing this innovation and giving their citizens the tools to take control of their financial futures.

However, not all nations are on board with this revolution. Countries like China and Russia have attempted to curtail blockchain innovation. China, for example, has created a cashless society with the launch of the Digital Yuan, a state-backed digital currency that has been in development since 2014. Meanwhile, Russia has proposed a ban on cryptocurrency mining and trading, with President Vladimir Putin expressing concerns about decentralized currencies' potential impact on the Russian economy. While these actions may hinder innovation within these countries, blockchain is continuing to thrive in many other parts of the world, and its growth is unstoppable.

As blockchain continues to evolve, it empowers individuals by giving them control over their finances, a degree of privacy, and freedom from centralized authorities. While some countries like Russia and China have attempted to regulate or block cryptocurrency use, blockchain's decentralized nature ensures that it can't be easily shut down. This is the power that blockchain offers—not just corporations and governments, but to individuals like you and me. It's a pathway to financial freedom, control, and confidence in the digital economy.

Major banks such as JP Morgan and Bank of America are already exploring blockchain technologies to enhance their services and remain competitive in the changing financial landscape. In fact, financial companies around the world are increasingly turning to blockchain solutions for secure and fast international payments. As of December 2023, over 300 financial institutions worldwide have adopted Ripple's technology to enhance their payment systems.

These partnerships span across all seven continents, with notable collaborations including:

- **South America**: The Bank of Colombia has tested its Central Bank Digital Currency (CBDC) on

  Ripple's platform.

- **Europe**: Ripple has partnered with Clear Junction to facilitate instant cross-border settlements between the UK and the European Union.
- **Asia**: Collaborations with SBI Holdings have enabled cross-border remittances across countries like Vietnam, the Philippines, and Indonesia.
- **Middle East**: Ripple has expanded its presence through partnerships with entities such as the Dubai International Financial Centre, aiming to advance blockchain adoption in the United Arab Emirates.

The adoption of blockchain technology is not just a trend—it's a paradigm shift that is shaping the future of money and industries worldwide. As the world continues to embrace this digital revolution, those who understand how to leverage blockchain will be positioned to thrive in the new economy.

In today's evolving financial landscape, blockchain technology is beginning to disrupt traditional banking systems, providing new avenues for secure transactions across the globe. One of the most exciting prospects on the horizon is the potential for blockchain to improve international payments, particularly in the realm of SWIFT transfers. For context, SWIFT (Society for Worldwide Interbank Financial Telecommunication) facilitates around $5 trillion in transactions daily. However, as we move into the digital age, blockchain presents an opportunity to replace or augment this aging system, offering faster, cheaper, and more transparent cross-border payments.

A key area of focus is the ISO 20022 standard, which is designed to standardize messaging protocols for financial institutions. ISO 20022 adoption could pave the way for blockchain-based cryptocurrencies to become integral parts of the international payment system. Several cryptocurrencies already show promise in adhering to this standard, and they are likely to play pivotal roles in the future of global finance.

The Five Cryptocurrencies Shaping the Future of Finance

Here are five cryptocurrencies that are well-positioned to align with ISO 20022 and potentially become the backbone of SWIFT payments:

1. **Ripple (XRP)**: Ripple, also known as XRP, is one of the most widely recognized blockchain networks. Over 300 financial institutions, including global giants like JPMorgan, Bank of America, and Western Union, utilize Ripple's network for fast and efficient cross-border transactions. However, Ripple was embroiled in a legal battle with the U.S. Securities and Exchange Commission (SEC), to attempt to classify XRP as a security. They lost the case, but despite this, Ripple's potential to revolutionize the global payments industry remains significant.

2. **Stellar (XLM)**: Stellar, represented by its native cryptocurrency XLM, is another blockchain network focused on enabling cross-border transactions. It aims to connect financial institutions, governments, and individuals through its decentralized network. Stellar's compatibility with ISO 20022 makes it a strong contender in the future of global payments.

3. **XDC (XDC Network)**: XDC, part of the XDC Network, is a hybrid blockchain designed for global trade and finance. It aims to simplify and optimize trade finance, with partnerships in the financial sector and the potential to adhere to ISO 20022.

4. **Algorand (ALGO)**: Algorand is a high-performance blockchain that enables fast, secure, and scalable decentralized applications. Its underlying technology is well-suited to meet the demands of financial systems, making it another cryptocurrency to watch in the ISO 20022 landscape.

5. **IBM Blockchain**: IBM's blockchain solution, developed in partnership with Stellar, is designed to facilitate secure and efficient enterprise transactions. With IBM's deep ties to the financial industry and a focus on compliance with international standards, IBM Blockchain could play a crucial role in reshaping the future of cross-border payments.

Other Promising Projects in the SWIFT Evolution

In addition to the five major players listed above, there are three more blockchain projects rumored to be on track to meet the ISO 20022 standard:

1. **Hedera Hashgraph (HBAR)**: Hedera Hashgraph is a fast and secure decentralized ledger platform that promises to offer the speed and scalability needed for enterprise use. HBAR is gaining traction as a potential candidate for SWIFT adoption.

2. **Quant (QNT)**: Quant is a blockchain interoperability platform that facilitates secure and seamless communication between different blockchains. It has been gaining attention in the finance sector as it could bridge traditional systems and newer blockchain networks.

3. **Cardano (ADA)**: Cardano is a blockchain platform focused on security, scalability, and sustainability. Known for its rigorous academic approach to development, Cardano is gaining momentum as a potential player in the future of decentralized finance (DeFi).

Blockchain's Impact Beyond Banking

While the integration of blockchain in banking and payments is groundbreaking, the technology's reach extends far beyond the financial industry. Blockchain is already transforming sectors like real estate, healthcare, and even the legal system. By leveraging decentralized systems and smart contracts, blockchain can streamline processes, reduce fraud, and provide greater transparency in various industries.

Blockchain is revolutionizing industries across the board by adopting Bitcoin as legal tender. A great example is El Salvador building a Bitcoin city at the base of the Conchagua Volcano, using geothermal energy to mine Bitcoin naturally. (We will get into mining Bitcoin later). This is a groundbreaking example of how blockchain technology can empower countries and their communities.

Another industry being impacted by blockchain is the gaming industry. Microsoft's acquisition of Activision Blizzard—creator of iconic games like Call of Duty, Overwatch, and World of Warcraft—demonstrates how blockchain and the metaverse are transforming the gaming landscape. As gaming platforms and virtual worlds continue to converge, we could see a shift in how players decide where to invest their time and money. With these major franchises potentially moving to Microsoft's ecosystem, gamers will have to weigh their options: stick with Sony, or switch to Microsoft for exclusive access to these massive games.

The Global Impact of Blockchain and Cryptocurrency

The world is witnessing the dawn of a new financial system, one that is decentralized, secure, and accessible to anyone, anywhere. Whether it's enabling financial inclusion in emerging markets or challenging traditional banking systems in developed economies, blockchain technology is set to change the way we think about money and financial transactions.

Blockchain takes a different approach than traditional banks, which lock your funds away, lend them out, and charge interest on your money while restricting how much you can send. Cryptocurrency does not charge you monthly fees; you can send as much as you want, and you can even create your own projects and applications on the blockchain. The opportunity is available to everyone, and it's incredible how opensourced these codes are for anyone who wants to seize the opportunity and join the space.

As we explore the potential of these technologies, their impact on global finance, and how they can be leveraged to secure your own financial future, realize that the path to understanding and utilizing blockchain may seem complex at first, but with the right guidance, you can navigate the space and harness its benefits.

Stay curious, do your own research, and embrace the possibilities that blockchain offers. Whether you're just starting out in the cryptocurrency space or already have experience in digital assets, this book is designed to help you unlock the full potential of the blockchain revolution.

Welcome to the Future of Finance—It's Just Beginning

The world of blockchain is rapidly changing how we think about business, finance, and innovation. It is now possible for anyone, from anywhere, to create their own cryptocurrency, attach it to a business as a utility, and turn it into a valuable project. This is one of the most remarkable aspects of blockchain technology: it has opened opportunities for entrepreneurs and developers to create new ideas and solve problems in ways that were previously unimaginable.

In blockchain, we often hear about terms like privacy coins, stablecoins, NFTs, staking, high yields, borrowing, lending, and wrapping. These are all powerful concepts that can be built on the blockchain, and many of them can be linked across different blockchains. As we explore these concepts, you'll begin to see how vast the world of blockchain really is. It's not just for the tech-savvy or investors; it's for every age group, culture, and country around the world. The opportunities are limitless for those who understand how to navigate the blockchain space.

One key to success in this space is the ability to research and evaluate projects effectively. This book will provide you with the tools and guidance you need to make informed decisions. But remember, as with any investment or venture, it's crucial to do your own due diligence. Research is your best friend, and understanding the projects you're interested in can make all the difference.

Blockchain is revolutionizing everything around us, especially as people are drawn into the Metaverse. As technology advances, so will ideas, and it's clear that as a society, we are shifting everything into the digital realm. From the way we use money, to how we communicate with one another; virtual reality is becoming increasingly central to our lives. It's something we once saw in movies like *Ready Player One*, where people leave their day jobs and earn millions by playing online games. The direction society is heading is truly mind-blowing.

In the chapters ahead, we'll delve deeper into the world of blockchain, cryptocurrency, and the decentralized economy. You'll learn how to leverage these emerging opportunities, understand the risks involved, and make informed decisions that can help you build wealth in the age of digital assets.

Chapter 2

What is Bitcoin and Why Does It Have So Much Value?

What is Bitcoin? Why does its price variate so much? Can you spend it like cash? Is it a scam? I don't even know what Bitcoin is! These are questions and statements we've probably all heard. The information you find when you search "bitcoin" online is often vague, fear filled or is so confusing to newcomers that they give up trying to understand.

I hear these exact questions when I speak with potential clients. I speak to over 100 clients daily, and there isn't a day that goes by that I do not come across someone who's never heard of Bitcoin. They don't know what it is, how it works, or how to buy it. Some potential clients have never even heard of the other cryptocurrencies—and there are over 10,000! This is one of the reasons I decided to create a crypto consulting business in 2016—to share the opportunities Bitcoin and blockchain technology had to offer.

In 2008, the United States went through a recession that shook traditional markets and severely affected the economy. In the wake of this, Bitcoin was created as a decentralized digital currency, officially launched in January 2009. It followed the principles outlined in the whitepaper by the mysterious pseudonymous figure Satoshi Nakamoto, the founder and creator of Bitcoin.

Bitcoin is built on the SHA-256 algorithm and uses a concept known as Proof of Work (PoW). In PoW, miners solve complex mathematical equations, called the nonce, to create the "genesis block," which is the first block in the Bitcoin blockchain. These miners compete in a process where they must solve these equations to validate Bitcoin transactions.

The first miner to solve the nonce gets a reward of 6.25 BTC for creating the next block. these mining pools, often established in large mining farms, are in constant competition to win the nonce and claim the Bitcoin reward.

As of now, miners are rewarded with 6.25 newly minted Bitcoins for each block they successfully mine, and at the time of writing this, each Bitcoin is worth about $68,000. This is one reason Bitcoin mining is so lucrative.

So, what drives the value of Bitcoin? One of the biggest factors contributing to Bitcoin's value is its finite supply of 21,000,000 coins. Unlike fiat currencies, which can be printed endlessly by governments and central banks, Bitcoin has a hard cap, meaning only 21 million Bitcoins will ever be mined. This limitation on supply contrasts with the inflationary nature of fiat currency, which, when printed excessively, causes inflation and reduces purchasing power.

The ongoing global economic challenges, worsened by supply chain disruptions from the COVID-19 pandemic, have worsened inflation. Many countries are experiencing severe currency devaluation, making it difficult for people to afford basic goods and services. In response to this, some nations, including El Salvador, Cuba, Switzerland, Ukraine, and Panama, have begun adopting Bitcoin as legal tender. These moves are part of a broader trend toward decentralization, where countries seek alternatives to their national currencies.

For example, in El Salvador Bitcoin mining is being innovatively done using natural resources, specifically by harnessing energy from volcanoes. This groundbreaking move by El Salvador could pave the way for more sustainable, eco-friendly Bitcoin mining methods.

Crypto Utopia and Using Volcanoes to Mine Bitcoin

One of the most pressing issues with Bitcoin mining today is the high energy consumption needed to power the Proof of Work (PoW) system. In fact, the New York Times published an article that highlights the environmental impact of Bitcoin's carbon footprint. When Bitcoin first launched in 2009, anyone with a simple desktop computer and processor could mine a single Bitcoin. The energy needed for mining back then was minimal, and in a matter of seconds, you could mine a coin. At that time, Bitcoin's value was only $0.06.

Fast forward to today, mining Bitcoin is far more complex and energy intensive. You need a room full of specialized hardware, such as GPU (Graphics Processing Unit) and ASIC (Application-Specific Integrated Circuit) mining rigs. Each rig can cost anywhere from $2,000 to $10,000, depending on the model. The energy consumption needed to mine Bitcoin today is equivalent to 9 years' worth of household electricity, costing about $12,500.

To put it into perspective, Bitcoin mining uses more energy than any of these countries use daily - Argentina, Finland, Chile, Denmark, the Netherlands, and Sweden. As a result, countries like China have started placing restrictions on mining operations causing miners to move to other regions like Kazakhstan, where coal-powered energy is cheap, or to areas in Texas, despite the state's struggling electric grid. Mining operations now often involve vast warehouses packed with powerful computers, working around the clock to solve complex mathematical equations, consuming massive amounts of energy in the process.

While the Proof of Work model has its merits, it also has significant drawbacks compared to Proof of Stake (PoS), which is an alternative consensus mechanism. PoW requires miners to use computational power to validate transactions and create new blocks, which is energy intensive. Later in this book, we will explore the pros and cons of both PoW and PoS to give you a better understanding of these two systems.

Out of Bitcoin's finite supply of only twenty-one million coins, it is estimated that not all bitcoins are still in circulation due to lost private keys (the 12–24-word backup password). If you lose your private key, you will lose access to your Bitcoin forever. Additionally, if you send Bitcoin to the wrong address, there is no way to reverse the transaction, which can be a serious risk. However, some cybersecurity firms specialize in recovering lost funds, but you need to act quickly to have any chance of recovery. The finite supply of Bitcoin contributes to its volatility, which is why we often see dramatic price swings. In basic economic terms, when the supply of an asset is limited, and demand increases, the price rises. This is one of the driving factors behind Bitcoin's price fluctuations.

Bitcoin is built on a secure algorithm that serves as a decentralized payment system on the blockchain, which acts as a ledger of trust. One of the most significant advantages of Bitcoin is its ability to prevent fraud and double spending. In traditional banking, a physical bill or coin can only exist in one place, so it cannot be spent twice. But in the digital world, it is possible to duplicate or "double spend" a digital asset, but with Bitcoin's peer-to-peer ledger, using the consensus mechanism ensures that once a transaction occurs, ownership is transferred without the risk of it being spent again.

Every Bitcoin transaction is recorded and publicly visible on the blockchain, which enhances transparency. With Bitcoin's halving event, the reward for miners solving the nonce (mathematical equation) decreases by half approximately every four years. The last halving occurred in April 2024, when the mining reward dropped from 6.25 BTC per block to 3.125 BTC per block.

Imagine winning the race to solve the nonce and create the next Bitcoin block. If you won the block right now, your reward would be worth approximately $425,000! This halving mechanism ensures that Bitcoin's supply remains controlled, and miners will continue to be incentivized to secure the network even as the reward decreases over time.

This combination of a limited supply, built-in scarcity, and secure decentralized network makes Bitcoin a revolutionary asset, not only as a store of value but also as a means of transaction in an increasingly digital world.

The competition in Bitcoin mining is fierce, and only one miner can win each new block created in the blockchain. If you want to compete with large mining pools, you'll need substantial upfront capital. Recently, Texas has been positioning itself as a major Bitcoin mining hub. With cheap land and a state government open to Bitcoin mining, we could be on the brink of a digital gold rush.

China recently (2025) lifted its ban on all crypto mining and trading. The Chinese Supreme Court had set heavy penalties for Bitcoin mining and trading, particularly for engaging in "illegal fundraising," which includes promoting or conducting cryptocurrency transactions via the internet, media, or other means. Offenders faced prison sentences of at least 10 years and fines of up to $79,100. Before China's crypto mining ban, the country accounted for 75.5% of the global Bitcoin mining power, generating an estimated $6 billion in crypto mining revenue. At that time, many miners moved to countries like Kazakhstan, where electricity is cheap, thanks to abundant coal reserves. China will likely gain momentum now that the ban is lifted.

In 2022, an internet shutdown in Kazakhstan, triggered by anti-regime protests, knocked out around 20% of the world's Bitcoin mining capacity. Kazakhstan had become a key destination for miners moving from China due to its cheap electricity costs. However, Kazakhstan's government has since sought to impose taxes on miners. This shows that many parts of the world are not happy with Bitcoin's enormous energy consumption, which has even led to power outages and internet disruptions.

Mindsets have changed recently with more countries beginning to open to cryptocurrency instead of shutting it down. The European Union, for example, has recently passed laws to allow cryptocurrency, especially as Russia's invasion of Ukraine has sent ripples through global markets. As a result of the conflict and international sanctions, the Russian ruble has lost significant value. Although this has created ways for Russia to avoid the sanctions and use cryptocurrency and other methods like gold.

Meanwhile, Ukraine has raised over $50 million in cryptocurrency donations, including Bitcoin and Polkadot, from various countries. Cryptocurrencies are playing a key role in helping Ukrainians flee the country safely, as they provide a means of accessing funds without relying on traditional banking systems.

And as sanctions on Russia continue to disrupt global markets, many individuals are turning to cryptocurrencies to protect their wealth. This has led to a surge in Bitcoin purchases, even at prices as high as $20,000 per coin, as people in Russia and other affected countries seek refuge from the destabilizing effects of currency devaluation and sanctions.

Banks' interest rates have surged by 20% making Bitcoin a welcome haven for many. In countries like Ukraine, where the financial system is heavily affected by geopolitical instability, people are turning to Bitcoin and other cryptocurrencies as a source of value.

Bitcoin has been a critical lifeline for Ukrainians, allowing them to buy armor, weapons, and essential military supplies, including tanks and helicopters, through cryptocurrency donations. Over $100 million has been raised for Ukraine's defense efforts, according to Alex Bornyakov, Ukraine's deputy minister for digital transformation and the government's crypto spokesperson.

Bitcoin has proven to be a game-changer, not only for Ukraine but also for the broader cryptocurrency ecosystem. Through blockchain technology, the possibilities seem limitless. From using Bitcoin ATMs to buy crypto to obtaining crypto-powered debit and credit cards, or even buying real estate and vehicles with bitcoin, the world of cryptocurrency is rapidly evolving.

I have experienced the power of cryptocurrency firsthand. I have been able to pay for houses, cars, travel, and daily essentials all through Bitcoin and other cryptocurrencies. There is no feeling quite like having the freedom to make anonymous transactions without interference from governments or central banks. This ability to spend crypto as easily as traditional money—even through a crypto debit card backed by VISA or MasterCard—shows just how far we have come. Later in this book, we will dive into the world of crypto debit and credit cards, highlighting some of the benefits and places where you can use them. Bitcoin is much more than just a digital asset—it is a payment system, often referred to as "Digital Gold," and it deserves to be viewed as such.

Chapter 3

Satoshi Value Simplified: The New Metrics of Money

Have you been confused when you see Bitcoin prices written in decimal form like this: 0.6799700? In this section, we are going to break down the metric system on which Bitcoin is built. Specifically, we will discuss Satoshi value. For some, this can be confusing, but it is beneficial to understand how the numbers are calculated even though one can easily just do a search on something like "$20,000 USD to BTC", and you will get the current exchange rate.

What is Satoshi Value and Where Did It Come From?

The term "Satoshi" is named after the mysterious creator of Bitcoin, Satoshi Nakamoto, who was mentioned in a previous chapter. In his white paper, entitled "Bitcoin: A Peer-to-Peer Electronic Cash System," he outlined the idea of a decentralized peer-to-peer network to solve the problem of doublespending, a problem seen in earlier cryptocurrency concepts. As of today, the Bitcoin that Satoshi Nakamoto holds is valued at over fifty-six billion dollars and has never been moved since Bitcoin first went online in 2009.

Satoshi value represents the smallest unit of Bitcoin. One Bitcoin (BTC) can be broken down into one hundred million Satoshi. This allows you to own fractions of a Bitcoin, which is great because, as you will see, Bitcoin's value can be quite high, and buying a full Bitcoin isn't necessary to get started.

Breaking Down Bitcoin's Structure

Bitcoin is divided into several smaller units. Here is how the Bitcoin unit structure works:

- 1 bitcoin (BTC) = 1,000 millibitcoins (mBTC)
- 1,000,000 microbitcoins (μBTC)

- 100,000,000 Satoshi (sat)

For example, if one Bitcoin is trading at $30,000 USD, then one Satoshi is worth approximately $0.00000030. It's a very small amount, which is why Satoshi is ideal for smaller transactions.

Satoshi is perfect for microtransactions, where you don't need to use large sums of Bitcoin. In fact,

many cryptocurrency exchanges offer the option to trade Bitcoin in Satoshi, especially when trading pairs are crypto-to-crypto, rather than crypto-to-fiat.

How to Calculate Satoshi Value From USD

To calculate the value of a certain amount of Bitcoin in Satoshi, you can use the following formula:

- Amount in USD ÷ Bitcoin Price (in USD) = Amount in BTC
- Amount in BTC × 100,000,000 = Amount in Satoshi

Next, to calculate the Satoshi value of a specific altcoin, you can use the following simplified equation:

- Satoshi value of altcoin= (100,000,000 Current price of Bitcoin) ×Specific altcoin price in USD\
- Satoshi value of altcoin} = 100,000,000 {Current price of Bitcoin {Specific altcoin price in USD}

Now, let's define these terms more clearly:

- Satoshi Price (s): The Satoshi value for a given altcoin.
- Current Price of Bitcoin (B): The current market price of Bitcoin in USD.
- Specific Altcoin Price in USD (Z): The price of the altcoin you are analyzing in USD.

Thus, the formula becomes: Satoshi Price (s)=(100,000,000B) ×Z\ {Satoshi Price (s) (100,000,000} {B} times Z

Example Calculation

Let's walk through an example to illustrate this formula clearly. Suppose the current price of Bitcoin (B) is $40,460 and the price of an altcoin (Z) is $50 USD. Now, apply our formula:

Satoshi Price= (100,000,00040,460) ×50=1,235.72 Satoshi per altcoin\ Satoshi Price} = {100,000,000} {40,460} \times 50 = 1,235.72 {Satoshi per altcoin}

So, the Satoshi value for one $50 altcoin at a Bitcoin price of $40,460 would be 1,235.72 Satoshi.

This formula can be applied to any cryptocurrency to calculate its Satoshi value. Since Bitcoin's market price (B) fluctuates rapidly, it's essential to remember that the Satoshi value of altcoins will also change accordingly, based on Bitcoin's price.

If you're not confident in doing the calculations manually, many cryptocurrency exchanges, such as Binance, offer tools that automatically handle conversions for you, saving you time and effort.

Now, let's go through an example to better illustrate how this works. Let's use an example to illustrate:

Below, you will see the pairs WABI/BTC and BTC/USDT (Image sourced from Binance).

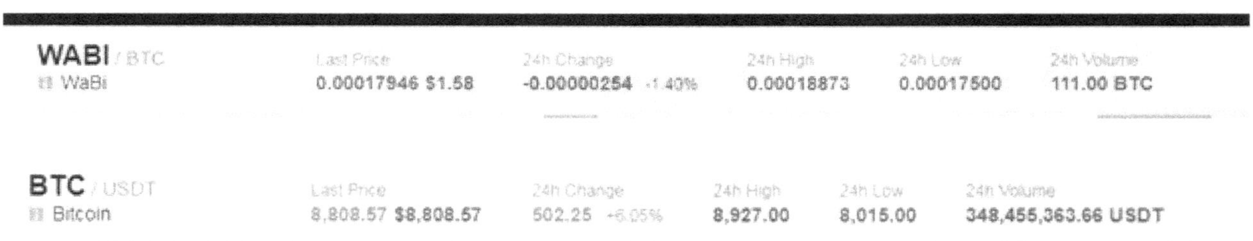

Pair	Last Price	24h Change	24h High	24h Low	24h Volume
WABI / BTC WaBi	0.00017946 $1.58	-0.00000254 -1.40%	0.00018873	0.00017500	111.00 BTC
BTC / USDT Bitcoin	8,808.57 $8,808.57	502.25 +6.05%	8,927.00	8,015.00	348,455,363.66 USDT

Let's walk through our formula step by step: (100,000,000 / B) x Z = s

For example: (100,000,000 / 8,808.57) x 1.58 = 17,937 Satoshi. It's as simple as that.

Tip: You can get a close approximation by multiplying the last price of an altcoin (like WABI) by 100 million. For instance, if the price of WABI is 0.00017946, then multiplying by 100,000,000 gives you 17,946 Satoshi, which is very close to the value we calculated above.

When tracking your profits, losses, and overall performance in cryptocurrency trading, the base currency you choose is important. Many exchanges don't offer USD equivalents, which can make it difficult to track performance, especially if you're using multiple exchanges. USD equivalents are only helpful if you're moving your holdings back to Tether (USDT); a stablecoin pegged to the US dollar, after each trade.

If you're using USD to track your gains, you'll be exposed to two currency pairs:

1. The specific altcoin value relative to Bitcoin (e.g., WABI/BTC).
2. Bitcoin value relative to Tether (USDT) (e.g., BTC/USDT).

In this scenario, until you trade your altcoin for Bitcoin (BTC) and then Bitcoin for USDT, you are essentially exposed to Bitcoin performance. Many traders ignore USDT when they're not actively trading, preferring to hold Bitcoin (BTC) or Ethereum (ETH) during that time.

Thus, understanding Satoshi value becomes crucial for tracking your portfolio's true performance. By learning this formula, you'll be able to make more precise decisions when buying, trading, and managing your portfolio.

How to Use Bitcoin and Satoshi in Real Life

Not all merchants accept Bitcoin or cryptocurrency, but there are many that do. When you're making a purchase or investment, always check whether the merchant accepts Bitcoin or Satoshi before continuing.

If you're using a cryptocurrency exchange, some will convert your Bitcoin or Satoshi into fiat currency for purchases, while others will keep it in cryptocurrency form.

Exchanges like Binance, KuCoin, Gemini, and Crypto.com allow you to trade Bitcoin in Satoshi form, especially when dealing with crypto-to-crypto pairs. Always make sure you understand the exchange rules and how Bitcoin is quoted, whether in Satoshi or in full BTC.

Chapter 4

What is Ethereum and Why is it Different than Bitcoin?

Ethereum is a decentralized, open source blockchain platform that revolutionized the cryptocurrency space. Created by programmer Vitalik Buterin in 2013 and officially launched on July 30, 2015, Ethereum has played a pivotal role in the development of the blockchain industry. The platform introduced concepts that have become fundamental to today's decentralized finance (DeFi) ecosystem, smart contracts, and decentralized applications (Dapps).

Ethereum's native cryptocurrency, Ether (ETH), is the second-largest cryptocurrency by market capitalization, following Bitcoin. Its starting price was just $0.31 per coin, and as of November 2021, ETH reached an all-time high of $4,891.70. The staggering rise in its value reflects the growing importance of Ethereum within the blockchain ecosystem.

The Innovation of Ethereum: Proof of Work and Proof of Stake

Ethereum, like Bitcoin, was initially built upon the Proof of Work (PoW) consensus algorithm, which is used to secure transactions and generate new coins through mining. However, Ethereum also introduced the concept of Proof of Stake (PoS), an alternative consensus mechanism designed to reduce energy consumption and increase scalability.

Ethereum's unique algorithm, Ethash, is a modified version of Proof of Work, combining both PoW and PoS elements. Ethash uses the Keccak-256 (SHA3-256) hashing function to decide the information that will be included in a block. This makes Ethereum more energy-efficient compared to Bitcoin's traditional PoW model, while keeping the security and decentralization that are critical to blockchain technology.

Smart Contracts and Decentralized Applications (Dapps)

One of Ethereum's most significant innovations was its implementation of smart contracts. Smart contracts are self-executing digital contracts that automatically perform the terms of an agreement between parties once predefined conditions are met. These contracts reduce the need for trusted intermediaries, lower transaction costs, and increase the reliability of digital agreements. Ethereum's ability to host these contracts on its blockchain has paved the way for a wide range of decentralized applications (Dapps) to be built on the Ethereum network.

The Ethereum blockchain functions like a "global computer," designed to be robust, censorshipresistant, and less prone to fraud. Co-founder Gavin Wood famously described Ethereum as a "one computer for the entire planet," capable of running programs on a globally distributed network of public nodes.

Ethereum and Tokens: ERC-20 Standard

Another major aspect of Ethereum is its ability to create and host other cryptocurrencies and digital

assets. These assets, known as "tokens," are often built using Ethereum's ERC-20 standard. Tokens are essentially smart contracts that run on the Ethereum blockchain, and they can stand for a variety of assets, from digital goods to real-world objects.

Since Ethereum's launch, more than 442,647 ERC-20 tokens have been created, and over 40 of them are ranked among the top 100 cryptocurrencies by market capitalization. ERC-20 tokens have become one of the most widely used standards for launching new cryptocurrencies, thanks to Ethereum's highly adaptable and decentralized platform.

The Ethereum Ecosystem: A Force Behind Cryptocurrency Adoption

Ethereum has played a crucial role in the broader adoption of blockchain and cryptocurrency. Its ability to host smart contracts and decentralized applications has fostered an entire ecosystem of new digital assets, platforms, and services. The Ethereum network has enabled projects like decentralized finance (DeFi), non-fungible tokens (NFTs), and more to flourish.

The flexibility and scalability of Ethereum have made it the backbone of numerous projects, allowing for the creation of everything from digital art marketplaces to decentralized exchanges and lending protocols.

In conclusion, Ethereum has transformed the cryptocurrency landscape by introducing groundbreaking technologies like smart contracts and Dapps. As blockchain space continues to evolve, Ethereum stays at the forefront of innovation, driving the next generation of decentralized applications and digital assets.

Key Points on Ethereum and ERC-20 Tokens

- Ethereum, one of the most popular cryptocurrencies and blockchain systems, is built around the concept of tokens, which can be bought, sold, and traded.
- A key innovation within Ethereum's ecosystem is the ERC-20 token standard. This technical framework defines the rules for how tokens are created, transferred, and interacted with on the Ethereum blockchain.
- Many well-known digital currencies and projects use the ERC-20 standard, including Maker (MKR), Basic Attention Token (BAT), Augur (REP), and OmiseGO (OMG).

Understanding ERC-20 Tokens

ERC-20 has become the definitive technical standard for smart contracts on the Ethereum blockchain, specifically for token implementation. This standard provides a set of rules that all tokens on the Ethereum network must follow, ensuring uniformity and interoperability. While ERC-20 tokens share many similarities with traditional cryptocurrencies like Bitcoin or Litecoin, they are distinct in one major way:

rather than using their own independent blockchain, ERC-20 tokens are issued and managed on the Ethereum network.

The rise of ERC-20 tokens has been monumental in the growth of the Ethereum ecosystem. As of October 2019, there were over 200,000 ERC-20-compatible tokens on Ethereum's main network. These tokens are not only digital assets that hold value, but they also enable various decentralized applications (Dapps) to thrive on the Ethereum blockchain.

The Significance of ERC-20 Rules

The ERC-20 standard defines a common set of rules that all Ethereum-based tokens must follow. These rules cover critical aspects of token functionality, such as:

- **Token Transfer Mechanism**: How tokens are transferred between users.
- **Transaction Approval Process**: How transactions involving the token are validated.
- **Token Access and Information**: How users can query information about the token.
- **Token Supply**: The total number of tokens in circulation.

By setting these guidelines, ERC-20 tokens become predictable and consistent, which significantly

benefits developers. Once a developer understands the ERC-20 standard, they can confidently create tokens that are compatible with Ethereum's vast network without needing to reinvent the wheel for each new project. This standardization reduces development time and increases reliability across the ecosystem.

Additionally, this consistency ensures that tokens built on Ethereum can seamlessly interact with one another. As the ERC-20 standard has gained widespread adoption, most tokens created through Ethereum's

Initial Coin Offerings (ICOs) are ERC-20 compliant, making them interoperable within the network.

Popular ERC-20 Tokens

Several well-known cryptocurrencies have been built using the ERC-20 standard, contributing to its widespread adoption. Some of these include:

- Maker (MKR)
- Basic Attention Token (BAT)
- Augur (REP)
- OmiseGO (OMG)

Because these tokens adhere to the ERC-20 standard, they can be easily traded and transferred across. Ethereum-compatible platforms and wallets.

ERC-20 and Wallet Compatibility

If you plan to buy or hold any ERC-20 tokens, it's essential to use a wallet that supports them. Fortunately, because ERC-20 tokens are so widely used, there are numerous wallet options available that ensure compatibility. This makes it easy for users to manage and store their tokens without needing specialized or unique wallets for each one.

The Six Essential Functions of ERC-20 Tokens

ERC-20 tokens are designed to ensure that they work cohesively within the Ethereum ecosystem. The standard includes six essential functions that dictate how tokens behave and interact with other parts of the network:

1. **transfer**: This function allows tokens to be sent from one address to another.
2. **transferFrom**: It enables the transfer of tokens on behalf of another user.
3. **approve**: This function allows a user to approve a third party to spend tokens from their account.
4. **allowance**: Users can check how many tokens an approved third party is allowed to spend on their behalf.
5. **balanceOf:** This function allows users to check the balance of tokens at any address.
6. **totalSupply:** This function returns the total number of tokens that exist.

These six functions ensure that all ERC-20 tokens behave uniformly, making it easier for both developers and users to interact with them. Whether you're transferring tokens, querying balances, or working with decentralized applications, these standardized functions provide the stability and reliability needed for a seamless experience.

The introduction of the ERC-20 standard has been a game-changer for the Ethereum network, giving rise to thousands of tokens and new decentralized projects. As the blockchain ecosystem continues to grow, ERC-20 tokens will remain a central component of Ethereum's success. With its standardized framework, the ERC-20 system has not only simplified token creation but also ensured the interoperability of tokens across the network, benefiting both developers and users alike.

Currency also supports ERC-20-compliant tokens. However, because the ERC-20 standard is still relatively nascent, there are likely to be bugs that need to be addressed. For instance, one notable past glitch involved Ethereum tokens sent directly to a smart contract. This error caused users to lose funds because the protocol design did not allow a token contract to respond to a direct transfer attempt, leading to the "loss" of funds associated with such transactions.

ERC-20 Tokens to Do Your Own Research (DYOR) On

Below are some of the ERC-20 tokens that are worth doing your own research on. Keep in mind that things change rapidly and what may have excelled at the time of writing this may not be as lucrative now.

- Chainlink
- USD Coin (USDC)
- Wrapped Bitcoin (1 BTC = 1 BTC on ETH)
- Binance USD (BUSD)
- Enjin Coin (ENJ)
- Status (SNT)
- Synthetix (SNX)
- Maker (MKR)
- Compound (COMP)
- Huobi Token (HT)
- Chiliz (CHZ)

- Nexo (NEXO)
- OKB
- UMA
- 0x (ZRX)
- Bancor (BNT)
- 1inch (1INCH)
- Nexus Mutual (NXM)

Expanding Ethereum Integrations

Ethereum's blockchain has fostered numerous integrations across different industries. Notably, the Play-to-Earn (P2E) gaming sector experienced a surge in interest during 2020 and 2021, driven by the rise of Non-Fungible Tokens (NFTs). NFTs have opened doors for unique in-game assets, collectibles, and the creation of decentralized economies, revolutionizing how users interact with digital assets.

Ethereum continues to evolve with advancements in decentralized finance (DeFi), scalable solutions like Layer 2 technologies, and increasing adoption in various sectors. Its ecosystem's growth showcases the innovation and adaptability of blockchain technology in real-world applications.

Another major factor in Ethereum's growth is the meteoric rise of NFTs, which has attracted millions of users to its blockchain. In January 2022, OpenSea, one of the leading NFT marketplaces, recorded a massive $5 billion in trading volume, marking a new all-time high. This surge in interest has been driven by high-profile sales and growing participation from both individual collectors and large companies. For example, the popular crypto game Axie Infinity alone has sold over $2 billion worth of Ethereum in NFT transactions. Well-known brands like Amazon, Microsoft, Ferrari, the NFL, and Gucci have all embraced NFTs, helping further establish Ethereum as the go-to blockchain for digital asset transactions.

Ethereum's role in these developments shows that it isn't just a cryptocurrency; it's a powerful ecosystem that facilitates a wide range of innovations, from staking and DeFi to NFTs and Play-to-Earn gaming. This growing versatility and adoption have cemented Ethereum's position as a fundamental player in the broader cryptocurrency and blockchain landscape, making it a valuable tool for both investors and developers alike.

Major global brands, including Ray-Ban, Original Penguin, McDonald's, Coca-Cola, Adidas, Louis Vuitton, Dolce & Gabbana, Nike, and many more, are diving into the world of NFTs, with the Ethereum blockchain serving as the foundation for these digital assets. NFTs created on the Ethereum blockchain typically use two specific token standards: ERC-721 and ERC-1155. These standards are designed specifically for NFTs, while ERC-20 tokens are reserved for cryptocurrencies that operate on the Ethereum network.

It's important to note that all Ethereum transactions are sent to the same Ethereum address. For instance, if your Ethereum address is: 0x1cCd7F0cda287d2Df5acCF19cBAc027929aC5FIFG, and you wish to send an NFT that is an ERC-1155 token, the process would involve copying your Ethereum address and pasting it into the designated send address. If you are on a platform like OpenSea, you would simply click "transfer" and enter your Ethereum address.

After completing this process, once the transaction is confirmed and signed, your ERC-1155 token will appear under ERC-1155 Token, where it can be tracked and managed with the corresponding visual representation. You can see this also if you go to https://etherscan.io/.

Companies are creating NFTs to establish exclusive, limited rare items and provide unique benefits such as access to discounts and events for owning their NFTs. This is one of the many advantages of using NFTs on the Ethereum blockchain.

Imagine selling a house: traditionally, you would go through a real estate agent and receive payment for the first sale. However, by attaching your house to a Smart Contract on the Ethereum blockchain, you could earn royalties every time the house is resold. For example, earning a 10% royalty on each subsequent sale would allow you to generate an ongoing income for the rest of your life, creating an entirely new revenue stream.

NFTs have opened endless possibilities for users to leverage networks like Ethereum, Solana (SOL), or Polygon to earn royalties. Attaching utilities or use cases to these NFT projects further incentivizes users to purchase them.

The Rise of Metaverse

Metaverse has emerged as one of the latest trends, with industries ranging from gaming to retail and even governments exploring virtual reality (VR) technologies like Oculus. Web 3.0 will form the foundation of Metaverse, consisting of blockchain-enabled decentralized applications that support an economy of userowned crypto assets and data. Ethereum plays a pivotal role in enabling these new technologies, offering developers and users countless opportunities to innovate and monetize.

Decentralized Exchanges (DEXs)

Ethereum's ecosystem includes decentralized exchanges (DEXs) such as Uniswap, SushiSwap, and 1inch. These platforms allow users to swap ETH for other ERC-20 tokens without transferring assets to centralized exchanges. By doing so, users only pay the network fees for the swap, avoiding additional charges.

DEXs offer significant advantages, including maintaining anonymity. Unlike centralized exchanges, which require users to undergo Know Your Customer (KYC) procedures—involving sharing sensitive information such as social security numbers and income details—DEXs enable crypto-to-crypto transactions without such invasive requirements. This appeals to users seeking privacy and autonomy.

Additionally, centralized exchanges often facilitate fiat-to-crypto transactions, which are subject to government control. Governments and traditional financial powers can freeze accounts or ban users from accessing their services. In contrast, decentralized currencies and DEXs empower users with more control over their assets and transactions.

Ethereum, like Bitcoin, has revolutionized the financial landscape by providing decentralized alternatives that challenge traditional systems, creating new opportunities for innovation and financial independence.

In recent years, we've witnessed a monumental shift in how we perceive money, with cryptocurrency leading the way toward breaking free from traditional banking systems and reshaping our understanding of financial transactions. Take, for example, a simple wire transfer. Typically, sending money through a bank can take anywhere from 7 to 10 business days for verification and clearance. Even with the fastest processing, where some transfers might clear within 24 hours, the process remains cumbersome and reliant on centralized institutions. But cryptocurrency challenges this outdated framework. With digital currencies like Bitcoin and Ethereum, the need for third-party intermediaries is eliminated. Money can be transferred directly from one person to another, peer-to-peer, in a matter of seconds—without the oversight of a bank or payment provider. Transactions on the Bitcoin network can often be confirmed in under 10 minutes, while Ethereum transactions are even quicker, with funds typically appearing in an account in a matter of minutes. This stands in stark contrast to the multi-day delays often experienced with traditional ACH transfers, which can take anywhere from 3 to 5 days to complete.

Cryptocurrencies like Ethereum open doors for businesses, freelancers, and digital entrepreneurs to engage in transactions without the limitations imposed by centralized financial systems. Small businesses, in particular, stand to benefit immensely. By accepting cryptocurrency, they can avoid chargebacks—a rapidly growing problem in e-commerce that leads to significant losses for online sellers.

Cryptocurrency transactions are irreversible, meaning once funds are sent, they are received, eliminating the risk of fraudulent chargebacks that plague traditional banking systems. This gives business owners peace of mind, knowing that they won't lose their hard-earned money to fraudsters.

Beyond chargeback prevention, cryptocurrency also addresses issues like double-spending, which can be a problem with traditional digital payment systems. In a cryptocurrency network, once a transaction is confirmed, it cannot be duplicated, ensuring the integrity of the funds. This level of security and transparency offers a compelling case for why cryptocurrencies, particularly Bitcoin and Ethereum, are superior to traditional money systems.

Another compelling reason to embrace cryptocurrency is the privacy it affords. Unlike conventional banking systems, which are subject to government oversight and regulation, cryptocurrencies give individuals greater control over their own finances. You can truly be your own bank, free from the constraints of centralized financial institutions that often limit access to your money or impose unnecessary fees.

Bitcoin, in particular, serves as a powerful tool for global transactions. It offers a seamless way to make payments across international borders, enabling users to buy everything from everyday essentials to cars, property, and even access to ATMs. Its growing acceptance as both a currency and a payment processor in both online and brick-and-mortar stores shows its versatility and potential.

Ethereum, however, takes the concept of cryptocurrency to the next level. While Bitcoin primarily serves as a digital currency, Ethereum introduces the revolutionary concept of smart contracts. These self executing contracts enable users to automate and enforce agreements across a wide range of industries— from real estate and healthcare to automotive sales, cannabis, and even government operations.

Ethereum's unique architecture also allows developers to build decentralized applications (DApps) and create new tokens through its ERC-20, ERC-721, and ERC-1155 standards. This not only enhances scalability but also fosters a thriving ecosystem, driving innovation and expansion in the world of decentralized finance (DeFi).

As more businesses and developers harness the power of Ethereum's blockchain, we can expect its influence to grow, fostering a new era of financial autonomy and technological advancement. With each transaction, every new application, and every emerging cryptocurrency, we are witnessing the birth of a decentralized future—one where individuals and businesses alike can operate outside the constraints of traditional financial systems.

Ethereum has fundamentally transformed the cryptocurrency landscape by introducing a decentralized algorithm that empowers projects to create their own tokens, enabling opportunities like staking and NFTs. These tokens are seamlessly integrated into Ethereum's main net, benefiting from the security, scalability, and reliability of its blockchain. Ethereum's innovation with smart contracts has not only redefined what's possible within the world of cryptocurrency but has also set the stage for other digital currencies to adopt similar technologies. As the pioneer of smart contract functionality, Ethereum has created a blueprint that others now follow, making it a cryptocurrency that is far more than just a payment system.

While Bitcoin remains the gold standard for peer-to-peer transactions, Ethereum has expanded the horizons of what cryptocurrency can achieve. It has opened a realm of endless possibilities—from decentralized finance (DeFi) applications and decentralized autonomous organizations (DAOs) to digital art and NFTs. By combining the power of smart contracts with the flexibility of its blockchain, Ethereum offers a platform that fosters innovation and brings the promise of a decentralized future into tangible reality.

Ethereum doesn't just prove that it can be used for payments like Bitcoin—it has become a cornerstone for the entire cryptocurrency ecosystem, creating avenues for developers, entrepreneurs, and creators to build on its blockchain and push the boundaries of what digital currencies can do.

Chapter 5

POW and POS: What are These Systems Used for and What are the Benefits?

In the cryptocurrency world, the mechanisms that underpin how transactions are validated and added to the blockchain are critical. Two of the most common consensus algorithms used by projects are Proof of Work (PoW) and Proof of Stake (PoS). Understanding how each works and their differences is crucial, as most blockchain projects are built on one of these systems. But what exactly are PoW and PoS, and why does it matter? We will take a look at them both separately.

Understanding Proof of Work in Bitcoin

To understand Proof of Work (PoW), we'll focus on how it operates within the Bitcoin network. Bitcoin is a decentralized digital currency built on a distributed ledger known as a blockchain. This blockchain records every transaction made with Bitcoin, grouped into "blocks" and arranged in a sequential chain. The primary goal of the blockchain is to prevent users from spending the same Bitcoin more than once—what is known as the double-spending problem.

The blockchain is public and distributed; any changes or tampering with the blockchain would be immediately noticed and rejected by the rest of the network. The key to detecting tampering lies in hashes— unique cryptographic outputs that act as proof of work.

When any data (such as a set of Bitcoin transactions) is run through a hash function (Bitcoin uses SHA-256 encryption), it generates a unique string of numbers and letters. This string, called a hash, is always of the same length regardless of the size of the original data set. What makes this process secure is the "avalanche effect": even a tiny change in the data will result in a completely different hash, making tampering easy to detect.

However, hashes alone are not enough to secure the Bitcoin network. If generating a hash was too easy, anyone could create one, undermining the integrity of the blockchain. To turn hash generation into genuine "work," Bitcoin employs a difficulty adjustment. This ensures that a new block is added to the blockchain roughly every 10 minutes, no matter how many miners are working. The difficulty is calibrated by adjusting the target for the hash. The lower the target, the more difficult it is to generate a valid hash. In practical terms, this means that a valid hash must start with a specific number of zeros.

Special Considerations: The Role of the Nonce

Given that any input of data only produces one hash, how do miners ensure they generate a valid hash below the set target? The answer lies in the use of a nonce—an arbitrary number that miners add to the data before hashing it. The nonce, short for "number used once," is altered repeatedly until a valid hash is found. Once the hash meets the target (i.e., it starts with the required number of zeros), the miner broadcasts it to the network, and the block is added to the blockchain.

Mining in Bitcoin is not just a race—it's more like a lottery. Since, on average, a valid proof of work is found every 10 minutes, but no one knows in advance who will find it. To increase their chances, miners often pool resources together, forming mining pools. By doing so, they share both the computational work and the rewards. The rewards consist of transaction fees and, for a limited time, newly created Bitcoins (known as the block reward).

Why Proof of Work Matters

Proof of Work ensures the security and integrity of the Bitcoin blockchain. Because the process of mining is resource-intensive, it makes altering any part of the blockchain incredibly difficult. To tamper with a single block, an attacker would have to re-mine that block as well as all subsequent blocks—a nearly impossible task due to the immense computational effort required.

Proof of Work creates a trustless environment, allowing participants to exchange value without needing to trust a central authority. Its decentralized nature, combined with the difficulty and energy cost of mining, makes the Bitcoin network highly resistant to fraud and tampering.

Proof of Work: Preventing Network Monopoly

One of the key strengths of the Proof of Work (PoW) system is its ability to prevent any single user or group of users from monopolizing the network's computational power. This is because the machinery and energy required to complete the hash functions involved in mining are both expensive and resource intensive.

Hard Fork: A Key Concept in Mining Networks

An important concept to understand in the context of PoW is the hard fork. If a group of miners or within a network begins accepting an alternative proof of work or makes changes to the consensus rules, it can lead to a hard fork. This is a situation where the blockchain diverges into two separate chains, each following a separate set of rules. A hard fork can occur when there's a disagreement among the network participants about the validity of new blocks or when a major upgrade is implemented that requires a notable change to the protocol.

Example of Proof of Work in Action

To better understand how Proof of Work operates, let's look at a practical example. When a miner participates in the Bitcoin network, they must compute a hash that satisfies certain criteria. Specifically, the hash must have a minimum number of leading zeros to be considered valid. This is part of the difficulty level that ensures a block is mined approximately every 10 minutes.

For instance, consider block #429818, which was successfully mined on September 14, 2016. The hash for this block is: 000000000000000004dd3426129639082239efd583b5273b1bd75e8d78ff2e8d.

This hash is the product of a hashing function applied to the data in that block. The difficulty was set in such a way that the correct hash needed to begin with a specific number of leading zeros (in this case, 18 zeros).

The miner who successfully found this hash was rewarded with 12.5 BTC (Bitcoin). Notably, this block contains a record of 2,012 transactions involving just over 1,000 BTC. Importantly, the block also includes a reference to the previous block's header, linking the blocks together into a secure chain.

What makes the Proof of Work system so secure is its impossibility to alter any part of the block. For example, if someone were to attempt to change a single transaction amount—say, adjusting it by even 0.000001 BTC—the resulting hash would be completely different. This would immediately alert the network to the discrepancy, and the altered block would be rejected as fraudulent. This inherent security makes PoW extremely effective in preventing attacks such as double spending.

Notable Proof of Work Coins

While Bitcoin is the most well-known Proof of Work cryptocurrency, it is by no means the only one. Numerous other cryptocurrencies also use the PoW consensus mechanism, each offering unique features or improvements. Here are some examples of popular PoW coins that are worth exploring:

- **Bitcoin (BTC):** Bitcoin is the pioneering cryptocurrency and remains the most well-known. It introduced the world to the concept of Proof of Work (PoW), establishing the fundamental model for securing decentralized networks. As the first blockchain to implement PoW, Bitcoin's mining process involves solving complex mathematical puzzles to add blocks to the blockchain, ensuring transaction integrity and network security.

- **Ethereum (ETH):** Ethereum, like Bitcoin, originally used PoW as its consensus mechanism. However, the Ethereum team has been transitioning the network to Proof of Stake (PoS) with the launch of Ethereum 2.0. The switch to PoS aims to reduce Ethereum's energy consumption and increase network scalability, though the transition has faced delays and challenges. Ethereum's PoW model, while still active, is being phased out in favor of PoS as the network evolves.
- **Litecoin (LTC):** Litecoin, often referred to as "the silver to Bitcoin's gold," is a direct descendant of Bitcoin and operates on the same Proof of Work mechanism. Litecoin was designed to offer faster block generation times and a more accessible mining process. Like Bitcoin, it relies on PoW to validate transactions and secure the network.
- **Bitcoin Cash (BCH):** Bitcoin Cash was created because of a hard fork from Bitcoin, with the goal of increasing transaction speed and reducing fees. As a PoW cryptocurrency, Bitcoin Cash follows a similar mining model to Bitcoin, requiring miners to validate transactions and secure the network by solving cryptographic puzzles.
- **Monero (XMR):** Monero is a privacy-focused cryptocurrency that uses Proof of Work to secure its network and validate transactions. Unlike Bitcoin and Ethereum, Monero employs a unique RandomX mining algorithm, which is optimized for general-purpose CPUs, making it more accessible for individuals with personal computing power to mine. Monero's emphasis on privacy and security, combined with its PoW consensus, has made it a popular choice among privacy-conscious users.
- **ZCash (ZEC):** ZCash is another privacy-oriented cryptocurrency that leverages zk-SNARKs (Zero-Knowledge Succinct Non-Interactive Arguments of Knowledge) to provide private transactions. Despite its focus on privacy, ZCash still uses a Proof of Work consensus model

to secure the network and validate transactions. The PoW mechanism in ZCash is like Bitcoin's, ensuring decentralized validation and the immutability of the blockchain.

- **Ethereum Classic (ETC):** Ethereum Classic is a continuation of the original Ethereum blockchain, followig a hard fork that occurred in 2016 after a contentious split within the community. Unlike Ethereum, Ethereum Classic continues to use Proof of Work and has no current plans to transition to Proof of Stake. It remains a separate project, but still shares many of Ethereum's features, including the ability to execute smart contracts.

Each of these coins, and others that use Proof of Work, have its own unique characteristics and can be valuable in different contexts, whether for investments, transactions, or technical exploration.

The Cost and Responsibility of Mining

Proof of Work (PoW) is widely used in cryptocurrency mining to validate transactions, secure the network, and create new tokens. However, PoW mining is resource-intensive, requiring significant computational power and energy.

The cost of mining equipment varies significantly, with specialized hardware called ASICs (Application-Specific Integrated Circuits) being the most powerful but also the most expensive. A single ASIC miner can cost anywhere from $7,000 to $11,000 or more, depending on the model and its processing capabilities. Additionally, miners need to cover the substantial operating costs of running these machines. For example, the cost to mine a single Bitcoin can reach around $34,000 when factoring in the equipment, electricity, cooling, and maintenance costs.

On top of these operational expenses, miners must also consider the transaction fees they incur, which can add an additional $40,000 or more in costs, especially for mining on high traffic blockchains like Bitcoin.

Despite these high costs, mining remains a critical component of many cryptocurrencies. Miners are responsible for several key functions in the blockchain ecosystem:

1. **Validating Transactions**: Miners check the validity of transactions by ensuring that they adhere to the network's rules. This includes verifying that the sender has enough balance to make the transaction and that it hasn't already been spent.
2. **Constructing a Block**: Miners group valid transactions into a block and try to solve a cryptographic puzzle (the Proof of Work) to add the block to the blockchain.
3. **Voting on Consensus**: By mining blocks and adding them to the chain, miners participate in the consensus mechanism, which determines the state of the ledger. This voting process ensures that only valid transactions are included in the blockchain.
4. **Broadcasting the Blockchain**: Once a new block is added, miners broadcast the updated blockchain to the rest of the network. This ensures that all nodes in the system have an up-to-date version of the ledger.

To ensure the integrity of the network and maintain a secure and decentralized system, miners must follow these protocols meticulously. Without miners working together to validate transactions and secure the blockchain, the entire cryptocurrency ecosystem would be at risk.

Challenges and Issues with Proof of Work (PoW)

While Proof of Work (PoW) has proven to be an effective consensus mechanism, it brings with it several challenges, particularly related to mining difficulty and environmental impact.

1. **Mining Difficulty**: The difficulty of mining is directly tied to the number of miners in the network. Bitcoin's mining difficulty adjusts automatically to ensure that blocks are mined

approximately every 10 minutes. This global difficulty level contrasts with the specific share difficulty used by mining pools, which is customized to suit the pool's needs. Mining pools allow miners to combine their computing power to increase their chances of successfully mining a block, thereby distributing rewards more evenly among participants.

2. **Energy Consumption and Environmental Impact**: One of the major criticisms of PoW cryptocurrencies like Bitcoin is their massive energy consumption. As more miners join the network, the computational power required to mine Bitcoin increases, which in turn demands more energy.

3. **Comparing Bitcoin's Energy Usage**: The environmental concerns related to Bitcoin's energy consumption are stark. Some estimates show that Bitcoin mining consumes more power than the entire nations of Argentina or Finland combined. This has raised alarms among environmentalists and led to calls for more energy-efficient consensus mechanisms, such as Proof of Stake (PoS), which do not rely on the same intensive computational processes.

Proof of Work (PoW) and the Prevention of 51% Attacks

Despite the environmental challenges associated with PoW, it offers a crucial benefit—protection against 51% attacks. A 51% attack occurs when a single individual or group gains control of over 50% of the network's total hashing power, which can allow them to manipulate the blockchain by reversing transactions or double-spending coins. This is typically achieved by renting substantial amounts of hash power from third-party providers or deploying a vast network of mining rigs.

The decentralized nature of PoW makes this scenario difficult to execute. While a 51% attack is theoretically possible, the likelihood of it happening in a well-established PoW network, like Bitcoin, is extremely low. The sheer amount of computational power required to control more than half of the network's mining capacity is prohibitively expensive. Thus, PoW provides a high level of security and resilience against such attacks, ensuring the integrity and trustworthiness of the blockchain.

However, Proof of Stake (PoS)—another consensus mechanism that has gained popularity in recent years—does not rely on expensive mining hardware, such as ASICs and GPUs. PoS offers its own set of benefits, primarily by reducing the energy consumption associated with mining. By eliminating the need for massive computational power, PoS is seen as a more environmentally friendly alternative, offering greater scalability and faster transaction speeds. In PoS, validators are chosen based on the amount of cryptocurrency they hold and are willing to "stake" as collateral, rather than competing through energy intensive mining.

The Origins of Proof of Stake

Proof of Stake was conceived in 2012 by two developers, Scott Nadal and Sunny King, who were motivated by the growing environmental concerns surrounding Bitcoin and its energy-intensive PoW model. At that time, the daily electricity costs for Bitcoin mining were a staggering $150,000, a figure that has only increased with the growing popularity of mining and rising energy consumption. By 2017, Bitcoin mining's daily electricity consumption was estimated to cost $6.7 million, based on the average cost of $0.12 per kilowatt-hour and the estimated 56,209,833 KWH consumed by the Bitcoin network on October 13, 2017.

Recognizing the unsustainable trajectory of PoW, Nadal and King proposed an alternative method of securing blockchain networks—staking. Rather than relying on energy-hungry miners to validate transactions and add new blocks, they suggested a system where participants (or validators) could stake their coins as collateral, earning the right to validate transactions and add blocks to the chain. The more coins a user staked, the higher the likelihood they would be selected to participate in block validation and earn rewards. This approach, they argued, would alleviate the rising costs and environmental strain associated with PoW mining.

What Is Proof of Stake (PoS)?

Proof of Stake is a consensus mechanism designed to achieve distributed consensus in a more energy-efficient and cost-effective manner. In a PoS-based blockchain, validators are selected based on the number of coins they have staked in the network, rather than the computational power they contribute through mining. This makes the process of validating transactions far less resource-intensive compared to Proof of Work.

To participate as a validator in PoS, users must first "stake" a certain amount of cryptocurrency, often Ether (ETH) in the case of the Ethereum network. The process works as follows:

1. **Staking**: In PoS systems, participants can earn rewards through staking, where they lock up their cryptocurrency holdings to support network operations. Staking typically offers annual percentage yields (APY) over various time periods, ranging from weekly terms to six months or longer. The longer the staking period, the higher the potential rewards.
2. **Validator Selection**: The blockchain protocol randomly selects validators based on the amount of cryptocurrency they have staked, with the larger stake giving users a higher chance of being chosen.
3. **Block Validation**: The selected validator verifies transactions and proposes a new block to be added to the blockchain.
4. **Rewards**: In return for their work, validators receive transaction fees and sometimes additional tokens as a reward. If the validator acts maliciously or does not perform their duties, they risk losing part or all their staked coins.

This system eliminates the need for expensive mining hardware and drastically reduces the energy consumption required to secure the network. PoS not only makes blockchain networks more eco-friendly but also improves scalability by reducing bottlenecks that occur with PoW mining.

Benefits and Challenges of Proof of Stake

Let's explore the benefits and challenges of Proof of Stake for a better understanding.

Benefits:

- **Energy Efficiency**: PoS consumes significantly less energy compared to PoW because it doesn't require vast amounts of computational power to secure the network. This makes it a more environmentally friendly option for blockchain consensus.
- **Scalability**: With lower resource requirements, PoS systems are better suited for scaling up and handling a larger volume of transactions at a faster rate.
- **Decentralization**: By allowing anyone with coins to participate in the validation process (as long as they meet the staking requirements), PoS can foster greater decentralization in blockchain networks compared to PoW, which tends to centralize mining power in the hands of those who can afford the expensive hardware.

Challenges:

- **Security Concerns**: Although PoS is considered more energy-efficient, it can be vulnerable to attacks such as nothing-at-stake attacks, where validators have little to lose by validating multiple competing chains. This issue is mitigated by introducing penalties for malicious behavior and slashing rewards for dishonest validators.

- **Centralization Risk**: There is a risk that individuals or entities with large amounts of cryptocurrency could dominate the staking process, leading to centralization. However, this is being addressed in some PoS systems through mechanisms that limit the maximum amount a single user can stake.

PoS as the Future of Blockchain

The introduction of Proof of Stake as an alternative to Proof of Work has the potential to transform cryptocurrency landscape. By reducing energy consumption and enabling faster transaction processing, PoS could play a key role in the future scalability and sustainability of blockchain technology. It allows cryptocurrency networks to remain secure and efficient without the environmental and financial costs associated with traditional mining.

As more blockchain projects move towards PoS, they also enable greater participation from the community, giving token holders a direct stake in the governance of the network. Through staking, users can earn rewards,

How Ethereum's Proof-of-Stake System Works

Ethereum, one of the largest and most prominent blockchains, transitioned from Proof-of-Work (PoW) to Proof-of-Stake (PoS) with the launch of Ethereum 2.0. This shift aims to address some of the major challenges faced by PoW, such as high energy consumption and scalability. Here's how Ethereum's PoS system works:

1. **Staking**: To participate as a validator, users must lock up 32 ETH (Ether) as collateral in the network. This staking process activates validators and ensures they have a personal stake in maintaining the integrity of the network.

2. **Validator Selection**: Validators are selected randomly from those who have staked ETH. Unlike PoW, where miners compete to solve complex puzzles, PoS validators are chosen at random to propose and validate blocks. This method eliminates the need for extensive computational power, reducing energy consumption.

3. **Block Creation and Validation**: When selected, validators create new blocks and confirm blocks proposed by others. This process is called attesting. Attesting is essentially a form of approval, where the validator confirms that a particular block is valid and in line with the network's consensus.

4. **Incentives and Penalties**: Validators are rewarded for proposing new blocks and attesting to others. However, to encourage good behavior and network security, validators can lose a portion of their staked ETH for actions like going offline (failing to validate) or engaging in malicious activity (such as colluding with other validators to attack the network). If a validator behaves maliciously, they risk losing their entire staked amount as a penalty.

The Future of Proof of Stake

As the cryptocurrency ecosystem matures, Proof of Stake is expected to become the consensus mechanism of choice for many projects. Its advantages in energy efficiency, lower barriers to entry, and enhanced security make it an attractive alternative to Proof of Work. Moreover, PoS aligns more closely with the growing global emphasis on sustainability, offering a solution to the environmental concerns that have plagued Proof of Work systems.

In addition, PoS enables a more decentralized and democratized model of governance. Validators have a say in decision-making, and stakers are incentivized to maintain the health of the network to ensure the value of their stake. This creates a self-sustaining, community-driven environment that fosters innovation and scalability.

Ultimately, Proof of Stake offers a promising solution to the challenges faced by Proof of Work, paving the way for a more secure, efficient, and sustainable blockchain ecosystem. As blockchain technology continues to evolve, PoS will likely play a pivotal role in shaping the future of decentralized networks.

When cryptocurrency values fall, the value of holdings also decreases. This creates a strong incentive for majority stake owners to maintain a secure and reliable network, as their investments are directly tied to its stability.

Examples of PoS Coins

- Binance Coin (BNB)
- Cardano (ADA)
- Polkadot (DOT)
- Stellar (XLM)
- Solana (SOL)
- Neo (NEO)
- Cosmos (ATOM)

Examples of PoS Cryptocurrencies

- Solana (SOL)

- Cardano (ADA)
- Avalanche (AVAX)
- Polkadot (DOT)
- Cosmos (ATOM)
- Polygon (MATIC)
- Algorand (ALGO)
- Waves (WAVES)
- Tezos (XTZ)
- Thorchain (RUNE)
- Harmony (ONE)

Real-World Issues with PoS Networks

Solana, one of the largest blockchain networks, experienced instability in January 2021 due to excessive duplicate transactions. This caused significant disruption, leading to a drop of over 30% in Solana's value during that period. However, the network has since been recovering and improving its infrastructure.

Additionally, Solana's integration with NFT marketplaces, such as OpenSea and its native platform SolanaFloor, has attracted significant attention. Minting NFTs on Solana is far more cost-effective compared to Ethereum, where gas fees can range from $100 to $300 per NFT. In contrast, Solana's low transaction fees make it a more attractive option for creators and collectors.

Proof of Stake represents a significant step forward in the evolution of blockchain technology, offering scalability, energy efficiency, and broader accessibility. However, addressing challenges like centralization, monopolization, and security vulnerabilities remains critical for long-term success and sustainability.

Choosing to Invest in PoW or PoS

The Internet of Things (IoT) is revolutionizing the way people connect and interact with technology. Whether you choose to invest in Proof of Work (PoW) or Proof of Stake (PoS) cryptocurrencies, both present exciting opportunities.

For those seeking a reliable store of value, PoW remains a strong choice. These projects are designed to grow in value over time due to their decreasing supply and periodic halving of mining rewards. As mining rewards are cut in half, scarcity increases, driving up demand and, subsequently, the value of these assets.

On the other hand, PoS offers unique advantages, especially when it comes to staking—a topic we'll explore in more detail later in this book. Staking allows users to earn an annual percentage rate (APR) by locking up their cryptocurrency for a specific period. This approach functions similarly to a savings account, but instead of banks profiting from the interest, investors directly earn the rewards.

Both PoW and PoS have significantly shaped the cryptocurrency landscape. PoS, in particular, is gaining popularity because it is more cost-efficient and environmentally friendly compared to PoW.

Ethereum serves as a prominent example of this transition. Originally based on a PoW system, Ethereum plans to fully shift to a PoS consensus mechanism. This upgrade aims to reduce energy consumption by 99%, improve scalability, and potentially enable the network to handle up to 100,000 transactions per second. Proponents of Ethereum highlight the economic incentives PoS offers to maintain network integrity. Validators who attempt to validate bad transactions or blocks face a penalty known as "slashing," where their staked ether is burned (permanently removed by sending it to an inaccessible wallet address).

Vitalik Buterin, Ethereum's co-founder, asserts that PoS is more secure than PoW. While attacking a PoW blockchain requires controlling more than half of the network's computing power, attacking a PoS system would require ownership of more than half of the staked coins—a challenging but distinct barrier.

Ethereum's PoS mechanism is already operational on the Beacon Chain, which launched on December 1, 2020. To date, over 9.5 million ETH (valued at approximately $37 billion) have been staked, reflecting strong confidence in the network's future.

In summary, both PoW and PoS have their strengths and continue to play pivotal roles in shaping the blockchain and cryptocurrency industries. Understanding their differences and advantages is essential for making informed investment decisions in this rapidly evolving space.

Chapter 6

Centralized Exchanges and Decentralized Exchanges: How Do They Work?

What is a Centralized Exchange? How does it work? Is it better than a Decentralized Exchange? In this chapter, we will discuss the main differences between a Centralized Exchange (CEX) and a Decentralized Exchange (DEX). We will also provide resources for credible exchanges, ensuring you can buy cryptocurrency on both platforms with confidence and trust in their reputation.

What exactly is a Centralized Exchange, and how does it work? How can you set yourself up on these platforms to purchase cryptocurrency? Centralized Exchanges (CEX), or cryptocurrency exchanges, are online platforms used to buy and sell cryptocurrencies. They are the most common method investors use to trade and manage their cryptocurrency holdings. The term "centralized" refers to the fact that these exchanges are controlled and regulated by a central authority, often influenced by government regulations and financial institutions.

Think of a Centralized Exchange like a middleman. It facilitates transactions between buyers and sellers while charging fees for its services. For example, Coinbase is a well-known Centralized Exchange that helps users conduct cryptocurrency transactions securely.

Buyers and sellers trust these middlemen to manage their assets, much like customers trust a bank to hold and manage their money. However, this trust comes at a cost. Centralized Exchanges typically charge three types of fees:

1. **Deposit Fees**: Charged when moving funds into the exchange.

2. **Trading Fees**: Applied when buying or selling cryptocurrencies.

3. **Withdrawal Fees**: Incurred when transferring funds out of the exchange.

These fees cover the cost of security, monitoring, and operational services provided by the exchange. One of the key advantages of Centralized Exchanges is their user-friendly interfaces, which make it easier for newcomers to navigate the cryptocurrency market. Additionally, they often offer customer support and higher liquidity, enabling faster transactions.

The Importance of KYC (Know Your Customer)

One essential requirement for using centralized exchanges is completing the Know Your Customer (KYC) process. This is a standard verification procedure that most users have already experienced when opening a bank account, applying for a loan, or obtaining a credit card. KYC involves submitting personal information, including:

- Full name and date of birth
- Social Security Number (SSN)
- A government-issued ID (e.g., driver's license or passport)
- A selfie holding the ID and a note with the current date and a reference number

The KYC verification process can take anywhere from 24 to 72 hours, depending on the platform. While this can be a drawback for users eager to start trading immediately, it is a necessary step to ensure compliance with regulations and maintain platform security. Users may face initial purchasing limits until they establish trustworthiness over time.

FDIC Insurance on Centralized Exchanges

One significant advantage of centralized exchanges, such as Gemini, Binance.US, Crypto.com, eToro, Paxos, and Coinbase, is that USD funds held on these platforms are often FDIC insured. The Federal Deposit Insurance Corporation (FDIC) protects up to $250,000 per depositor in the event of bank failure. However, it's crucial to note that FDIC insurance only covers fiat (USD) balances on these exchanges. Cryptocurrency holdings, such as Bitcoin or Ethereum, are *not* FDIC insured.

Risks and Drawbacks of Centralized Exchanges

Despite their benefits, centralized exchanges come with notable risks and limitations. Users must understand that these platforms have the authority to:

- **Freeze Accounts**: They can suspend buy/sell services, effectively locking users out of their assets.
- **Restrict Withdrawals**: Users may lose access to their funds without prior notice.
- **Terminate Accounts**: Exchanges reserve the right to close accounts at any time without providing an explanation.

These risks mirror similar powers held by traditional banks. While such measures are often enacted for security, regulatory compliance, or suspicion of illegal activity, they can still result in frustration and economic loss for users.

Popular Centralized Exchanges

Newcomers to cryptocurrency are often drawn to centralized exchanges due to their user-friendly interfaces, clear instructions, and customer support. Popular centralized exchanges include Coinbase, Robinhood, Kraken, Gemini, Crypto.com, and Coinmama. These platforms simplify the onboarding process, which is essential for individuals unfamiliar with decentralized exchanges. Users rely on these platforms not only to securely facilitate transactions but also to leverage the exchange's extensive network of users to find suitable trading partners.

At present, it is more common for centralized exchanges to offer cryptocurrency-to-cryptocurrency (crypto/crypto) pairings. For example, customers can trade Bitcoin for Ethereum using a pair like BTC/ETH. In traditional financial markets, a similar pairing example would be USD/JPY (US Dollar/Japanese Yen). Fewer exchanges offer fiat-to-cryptocurrency pairs, such as Bitcoin for USD (BTC/USD). However, some of the largest exchanges, like Binance, dominate liquidity in both crypto/crypto and fiat/crypto pairs. These fiat pairs are crucial because they provide a direct entry point for new investors into the cryptocurrency market.

In conclusion, centralized exchanges remain a popular entry point into the cryptocurrency market due to their convenience, liquidity, and regulatory safeguards. However, users must weigh these benefits against the risks and limitations, especially regarding asset control and platform authority. As we move forward, we'll explore how decentralized exchanges (DEXs) address many of the challenges posed by centralized exchanges (CEXs) while introducing their own unique set of opportunities and risks.

One of the most significant differences is control over funds. When you deposit your money into a centralized exchange, although it remains technically yours, the exchange assumes control over it. This power dynamic allows centralized exchanges to freeze accounts, withhold funds, or even terminate user accounts without explanation. These are the very problems that decentralized exchanges aim to solve.

Decentralized Exchanges

The concept of decentralized exchanges began gaining traction on January 3, 2014, with the rise of DeFi (Decentralized Finance) applications. Powered by Automated Market Maker (AMM) systems, DEXs have evolved significantly, overcoming limitations of earlier order book models. A decentralized exchange allows users to conduct direct peer-to-peer cryptocurrency transactions over secure online platforms without the need for intermediaries, such as banks or centralized trading platforms.

In a typical centralized exchange transaction, a third-party entity oversees custody and transfer of assets. In contrast, decentralized exchanges eliminate this intermediary role, replacing it with blockchain technology and distributed ledgers. This approach significantly reduces risks associated with single points of failure, manipulation, or corruption. By using smart contracts, DEXs autonomously execute transactions, ensuring transparency, security, and efficiency.

Key Benefits of Decentralized Exchanges

1. **Peer-to-Peer Transactions**: Users can trade directly with each other without relying on intermediaries.
2. **Lower Fees**: Typically, users only pay network fees, which are often lower than the fees on centralized exchanges.
3. **No Account Freezes or Terminations**: Since there is no central authority, accounts cannot be arbitrarily frozen or terminated.
4. **Full Control of Funds**: Users maintain control of their assets and private keys, offering enhanced security.
5. **Instant Access**: Users can start trading immediately without waiting for lengthy account verification processes.

However, it's important to note that decentralized exchanges are not FDIC-insured. Even on centralized exchanges, FDIC insurance only applies to USD funds held in fiat accounts, not to cryptocurrencies themselves.

The Role of Smart Contracts

DEXs rely heavily on self-executing smart contracts to facilitate trading. These contracts automate processes, minimize the need for human intervention, and enable instant trades, often at lower costs. In a non-custodial framework, users are solely responsible for managing their cryptocurrency wallets and private keys. While this control is empowering, it also introduces risks:

- If private keys are lost, stolen, or destroyed, assets become permanently inaccessible.
- In cases of incapacitation or sudden death, if no one has access to the keys or backup documentation, the assets remain locked forever.

To mitigate these risks, it's essential to:

- Write down your private keys on paper (at least 2-3 copies).
- Store these copies securely in different physical locations.
- For those with significant crypto assets, consider including wallet access details in a legally documented will.

Regulatory Differences of DEXs

Unlike centralized exchanges, most DEXs operate outside the scope of Know-Your-Customer (KYC) and Anti-Money-Laundering (AML) regulations. While this enhances user privacy, it can also create vulnerabilities, such as increased potential for illicit activities. However, for users prioritizing anonymity and independence, DEXs remain an attractive option.

Limitations of DEXs

1. **No Fiat Integration**: DEXs only support crypto-to-crypto pairs, unlike CEXs, which enable fiatto-crypto transactions.
2. **Technical Complexity**: Managing wallets and private keys can be challenging for beginners.
3. **Smart Contract Risks**: Bugs or vulnerabilities in smart contracts can result in monetary loss.

4. **Limited Asset Availability**: New tokens often launch on centralized exchanges before appearing on DEXs.

Order Books and Swaps

Decentralized cryptocurrency exchanges (DEXs) and DeFi products have revolutionized the way users interact with digital assets. The first generation of DEXs relied on order books, similar to conventional centralized exchanges (CEXs). These order books maintain a record of all open buy and sell orders for specific assets, and the difference between these prices determines the order book's depth and the prevailing market price. On DEXs utilizing order books, transaction details are typically stored on-chain, while user funds remain off-chain in their wallets, ensuring both transparency and user control.

Some DEXs that still utilize the order book model include Serum, dYdX, Nash Exchange, TomoDEX, ViteX, Loopring Exchange, and Binance DEX. These platforms provide traders with flexibility and a traditional market experience while leveraging blockchain technology for transparency and security.

However, the new generation of DEXs has moved away from order books, instead using liquidity pool protocols to facilitate trades and determine asset pricing. These peer-to-peer platforms execute trades directly between users' wallets through mechanisms like Automated Market Makers (AMMs). This process, often referred to as a "swap," allows for instant transactions without requiring an intermediary.

These liquidity pool-based DEXs are typically ranked by Total Value Locked (TVL), which represents the total value of assets held in the protocol's smart contracts. Some of the most reputable and widely used DEXs in this category include Uniswap, Curve, SushiSwap, DoDo, 1inch, Balancer, SpookySwap, Phantom, Bancor, Kyber, AirSwap, Anyswap, IDEX, PancakeSwap, Raydium, and Trader

Joe. These platforms have earned credibility through consistent performance, security, and user satisfaction.

It's worth noting that certain DEXs are network specific. For example, Uniswap, SushiSwap, and 1inch primarily use Ethereum-based pairs for trading. However, platforms like Uniswap have recently expanded to other networks, including Polygon, Optimism, and Arbitrum, allowing users to trade assets across different blockchain ecosystems.

In summary, decentralized exchanges offer an innovative alternative to centralized platforms, with significant advantages in terms of control, transparency, and reduced reliance on intermediaries. However, they also have their own set of challenges, particularly around wallet management and regulatory compliance. Understanding these differences is crucial for anyone looking to navigate the cryptocurrency ecosystem effectively.

Centralized Exchanges vs. Decentralized Exchanges

When choosing between a centralized exchange (CEX) and a decentralized exchange (DEX), the fundamental difference lies in control and autonomy over your funds. With centralized exchanges, while they offer convenience, fiat on-ramps, and customer support, they ultimately retain control over your assets. This means your funds can be frozen, your account terminated, or your access restricted based on their guidelines, regulatory pressures, or internal policies.

In contrast, decentralized exchanges provide you with complete ownership and control over your funds. DEX platforms enable users to execute atomic swaps and transactions anonymously, eliminating intermediaries and significantly reducing counterparty risks. This level of independence is a core principle of decentralized finance (DeFi) and is one of the key reasons many users prefer DEX platforms.

If you find that a specific cryptocurrency pair is only available on a centralized exchange, the best practice is to purchase the asset and then transfer it immediately to a hardware wallet. Hardware wallets, also known as cold storage, keep your funds offline, making them significantly more secure against hacks, breaches, or unauthorized access.

Ultimately, peace of mind comes from knowing that your funds are under your control, safeguarded from restrictions or unforeseen limitations imposed by third parties. Whenever possible, prioritize using decentralized exchanges for trading and asset management. This approach ensures you retain sovereignty over your assets and reduces the risks associated with centralized control. In the world of cryptocurrency, self-custody and decentralization embody the true spirit of financial freedom.

Chapter 7

What is DeFi and How Does It Break Centralized Finance?

So, how does DeFi work? Decentralized finance (DeFi) uses blockchain technology, the foundation of cryptocurrencies. A blockchain is a distributed and secured database or ledger on which decentralized applications (dApps) are built. These dApps handle transactions, run blockchain operations, and enable services like staking. Staking is measured through a system called Total Value Locked (TVL), representing the sum of all cryptocurrencies staked, loaned, borrowed, deposited into pools, or used for other financial activities across DeFi. TVL can also represent the value of specific cryptocurrencies like Ethereum, YFI, COMP, MAKER, and others.

In a blockchain, transactions are recorded in blocks and verified by miners or validators in Proof of Stake systems. Once verified, the block is encrypted and linked to the next block, forming a chain. Each block contains information about the previous one, making it nearly impossible to alter historical data without affecting subsequent blocks. This structure, combined with security protocols, ensures the integrity and immutability of blockchain data.

Peer-to-peer (P2P) financial transactions are a core concept of DeFi. In a P2P transaction, two parties directly exchange cryptocurrency for goods or services without a central intermediary. For example, in centralized finance, securing a loan requires applying through a bank or lender, paying interest, and incurring service fees. In DeFi, you can use a dApp like NEXO to specify your loan needs. An algorithm matches you with suitable lenders, and after agreeing to their terms, you receive your loan—no lengthy approval processes, no excessive fees.

P2P lending on DeFi offers flexibility, allowing you to choose lenders globally. Transactions are recorded on the blockchain, ensuring transparency and consensus verification. Repayments are similarly tracked, maintaining a secure record of payments.

DeFi primarily operates using cryptocurrencies, including stablecoins—digital assets pegged to fiat currencies like the US dollar. A prominent example is USDT (Tether). Stablecoins are great at providing stability in the often-volatile crypto market.

If you're considering exploring or investing in DeFi tokens, here are some top projects by market capitalization: Terra, Avalanche, Wrapped Bitcoin, Dai, Chainlink, Uniswap, Fantom, Theta Network, Tezos,

THORChain, Maker, PancakeSwap, The Graph, Basic Attention Token, Loopring, Curve DAO Token, Convex Finance, Theta Fuel, Secret, Oasis Network, Compound, Yearn Finance, SXP, Synthetix, Ox, SushiSwap, Serum, and Injective.

These are just a few examples out of over 500 credible DeFi projects, with thousands more emerging as the space continues to grow.

Everyday new opportunities and new DeFi projects are being created. The future of Defi is bright and still is in the beginning stages of its evolution. So, since DeFi is largely unregulated, that means the ecosystem is still prone to infrastructural mishaps, hacks, and numerous scams.

Many existing financial laws were established by regulatory bodies such as the SEC and other financial authorities, each with its own set of rules and guidelines. For example, China recently banned cryptocurrencies and mining activities, and other countries like Egypt and Qatar have implemented similar restrictions. DeFi's borderless transaction capabilities raise significant questions for regulators worldwide.

Centralized authorities and governments recognize the power of DeFi and perceive it as a significant threat to traditional financial systems, as it disrupts their control and challenges the status quo. This is why I strongly encourage readers of this book to DYOR (Do Your Own Research) and explore the opportunities that DeFi presents. Take control of your financial future and become your own BANK!

Chapter 8

Privacy Coins Why Do I Need Them?

In this chapter, we will explore what privacy coins are and highlight some of the top projects worth researching. It's important to note that cryptocurrency projects evolve quickly, and while I'll share the ones I believe have long-term potential, this is *not financial advice*. Always *do your own research (DYOR)* to make informed decisions and avoid impulsive choices in the market.

Privacy coins are cryptocurrencies built on blockchain technology, typically with a finite supply, and are often issued by private organizations, corporations, or nonprofit entities. They act as an alternative to national or fiat currencies, which are traditionally the standard units of value in most countries. However, unlike fiat currencies, privacy coins are not legal tender in many jurisdictions and may even be outright illegal in some places. This lack of governmental control is one of the reasons governments often fear and attempt to regulate or ban them.

The Unique Value of Privacy Coins

One of the standout features of privacy coins is their resistance to inflation. While central banks can print fiat currency endlessly, devaluing their purchasing power, privacy coins have a *fixed supply*. This scarcity ensures they cannot be inflated, adding long-term value to their ecosystem. Privacy coins provide users with financial privacy, enabling secure and private transactions while still maintaining transparency and accountability through blockchain technology.

A Brief History of Private Currencies

The concept of private currencies isn't new. Private currencies date back to the 18th and 19th centuries during the Industrial Revolution in both the U.S. and Europe. During this time, mining towns and remote industrial hubs often issued their own forms of private currency, commonly referred to as *company scrip*. These currencies were often only redeemable at company-owned stores.

By the 1860s in the United States, there were over 8,000 different types of paper money in circulation. However, after the Civil War, the *National Banking Act* standardized the monetary system, eliminating these currencies and establishing a national currency backed by the federal government. Despite legal restrictions, private currencies persisted and continued to circulate in limited capacities.

Today, privacy coins have taken the place of traditional private currencies, leveraging blockchain technology to ensure secure and private transactions across global networks.

Technological Foundations of Privacy Coins

Privacy coins operate on different blockchain protocols to ensure transaction privacy. For example:

- **Proof of Work (PoW)**: Used by Monero, this system relies on miners to validate transactions while maintaining privacy through cryptographic techniques.
- **Proof of Stake (PoS)**: Used by Oxen, this system secures transactions while offering a more energy-efficient validation process.

Some privacy coins are also backed by physical commodities, such as gold or silver, further stabilizing their value and reducing susceptibility to inflation.

The Future of Privacy Coins

Despite overcoming significant technological and legal obstacles, privacy coins remain controversial. Many governments continue to tighten regulations around their use, fearing their potential to bypass financial controls and disrupt traditional banking systems. However, privacy coins have shown resilience, and some economists believe they could eventually become part of the mainstream financial system.

In the following sections, we will explore some of the top privacy coin projects worth researching. These projects represent innovation, security, and privacy in the evolving world of decentralized finance (DeFi). Stay tuned as we delve deeper into the technology and potential behind each of these coins.

Two Attributes of Privacy Coins: Anonymity and Untraceability

Privacy coins have the potential to replace some national currencies, and understanding their benefits can help clarify why individuals might want to own them. Privacy coins typically provide privacy in one of two ways: anonymity and untrace ability.

Anonymity focuses on hiding the identities of the people behind transactions. With privacy coins, people cannot see your name (e.g., Mike Rodgers) or your public address, and no one can trace your identity back to a public key. This means even powerful organizations such as the FBI or DEA cannot track you.

However, centralized cryptocurrency exchanges that require users to go through KYC/AML (Know Your Customer/Anti-Money Laundering) processes can track and report transactions to agencies such as the IRS and other federal authorities.

Untraceability, on the other hand, prevents people or computer systems from following a transaction trail. Privacy coins use several strategies to achieve these goals, with two of the most popular being CoinJoin and Coin Mixers (like the TOR Browser).

How Coin Mixers Work: Mixers combine multiple transactions into a single transaction and then divide that transaction to send the relevant amounts to each recipient. Essentially, the system mixes your cryptocurrency with a large pool of other cryptocurrencies and sends smaller units to an address of your choosing. The total amount sent is slightly reduced (typically by 1-3%) as a service fee for the mixing company. This process ensures that the cryptocurrency each recipient gets is from a combination of senders rather than a single identifiable source.

Who Benefits from Privacy Coins?

1. **Businesses**: Companies making large transactions in cryptocurrencies may not want these transactions to be traced. If competitors discover what they are buying, trading, or the quantities involved, it could pose significant business risks.

2. **High Net-Worth Individuals**: Wealthy crypto holders may want their transactions to remain secret to avoid becoming targets for hackers. For example, if a hacker learns that "Jason" recently transacted 30 Bitcoins (approximately $1,173,855), they might target him specifically.

3. **Enthusiasts, Escapists, and Radicals**: These groups often value increased financial privacy for ideological or practical reasons.

Anonymity for cryptocurrency transactions appeals to a wide range of individuals and groups. These include people who believe that governments or institutions should not have the ability to track every financial transaction. For such individuals, tools like coin mixers become highly attractive.

However, criminals are another group who value complete anonymity in cryptocurrency transactions. This association raises concerns within the cryptocurrency community, as it could lead to increased regulations on Bitcoin and other digital assets.

Maintaining Anonymity

Privacy coins utilize several methods to maintain anonymity, including Ring Addresses. Ring addresses connect multiple wallet addresses, making it nearly impossible to determine which address sent or received a transaction. Monero (XMR), with a total supply of 18,091,938, is one of the leading privacy coins employing this technique.

Another technique is Stealth Addresses, which creates a unique, one-time address for every cryptocurrency transaction. Monero also utilizes this method, specifically through the dual-key stealth address protocol, to enhance privacy.

Zero-Knowledge Succinct Non-Interactive Argument of Knowledge (zk-SNARK) is another advanced cryptographic system used by privacy coins. This protocol allows one party to prove a transaction's validity without revealing sensitive details like transaction amounts or wallet addresses. A helpful analogy is verifying your age without disclosing your exact birthdate.

Zcash (ZEC), with a total supply of 21,000,000, is a prominent cryptocurrency utilizing zk-SNARKs, providing optional shielding for transactions to ensure total anonymity.

Other cryptocurrencies, such as Loopring and Immutable X, leverage ZK-Rollups or zkSync to achieve similar privacy and efficiency goals.

Dash (DASH), launched in 2014 with a total supply of 18,900,000, offers users the option to enable transaction privacy through its PrivateSend feature. This functionality allows users to obscure the origins of their funds while maintaining compliance with their local regulatory standards.

DASH achieves this through a mixing protocol utilizing an innovative decentralized network of servers called master nodes. Horizen, with a total supply of 21,000,000, uses a different privacy feature than the others mentioned. Horizen offers privacy-shielded Z-Addresses and public T-Addresses that work similarly to Bitcoin. However, sending funds from a Z-Address to a T-Address will show the amount received. Horizen's vast node network is a benefit, helping to improve anonymity.

These innovative techniques and protocols demonstrate the significant strides privacy coins have made in safeguarding user anonymity while balancing regulatory concerns. As the technology continues to evolve, privacy coins will remain a critical part of the cryptocurrency landscape.

These are a few privacy coins that have implemented different systems to maintain anonymous transactions for both peer-to-peer interactions and corporations seeking privacy. I firmly believe that every man and woman in this world deserves the right to spend, buy, trade, or use their money as they please without government interference or tracking. True financial freedom means control over your own assets, not surveillance.

I genuinely believe privacy coins represent the future and align with the original vision of cryptocurrency: to break away from centralized financial systems. At the time of writing this book, here are the top Privacy Tokens by Market Capitalization. Please ensure you DYOR (Do Your Own Research) before purchasing any of these tokens:

- Monero
- Tornado Cash
- Zcash
- Secret
- Oasis Network
- Decred
- Horizen

- Keep Network
- Status
- Mobilecoin
- Pirate Chain
- Verge
- Dusk Network
- Phala Network
- Beam
- Firo
- Oxen
- Grin

To stay updated on the top privacy coins you can visit these websites to check the top coins and conduct thorough market research on projects:

- coinmarketcap.com
- coingecko.com
- coincheckup.com
- coinpaprika.com

Privacy coins like these offer significant benefits, including security, privacy, and fungibility. We will discuss these sites in greater depth later in the book.

Protecting Data Privacy

Data privacy has become a major issue, especially in the United States. The number of reported data breaches jumped 68 percent in 2023, reaching the highest total ever. A new report released in 2022 by the Identity Theft Resource Center highlights the growing concern over data security and the critical need for solutions that prioritize privacy.

According to the Identity Theft Resource Center's 2021 Data Breach Report, there were 1,862 data breaches, surpassing both 2020's total of 1,108 and the previous record of 1,506 set in 2017. With so many breaches affecting banks and major financial systems, cybersecurity has become an especially urgent concern in recent years. Cybercriminals, including hackers and scammers, now have highly advanced techniques and technology at their disposal, often using tools like TOR Browser, making data breaches increasingly common.

As a result, demand for platforms and programs that protect users' sensitive information from being tracked or stolen is at an all-time high. Naturally, people are particularly protective of their financial information. This is where the value of privacy coins becomes evident. Conventional cryptocurrencies like Bitcoin conduct transactions pseudonymously, but transaction details—such as amounts and wallet addresses—remain visible on the public ledger and can still be traced back to individual users without much difficulty.

The same issue exists with many cryptocurrencies, as their transaction histories and previous owners can be tracked on their blockchain's public ledger. Authorities can thus confiscate coins with a history of use in criminal transactions. Owners of these coins may also face interrogations or even jail time if found guilty, even if they themselves have never used cryptocurrencies for illicit purposes. Many users are understandably fearful of attracting attention from financial watchdogs, governments, and banking systems.

For such individuals, privacy coins represent a safer investment, as they are fully fungible and impossible to trace. Privacy coins offer users the opportunity to maintain full control over their financial transactions without interference. I hope this book empowers you to take control of your financial freedom, embrace privacy coins, and become your own bank today.

Chapter 9

Smart Contracts the Link to Broken Systems

In this chapter, we will dive deep into what smart contracts are, how they work, and their significance in the blockchain ecosystem. Smart contracts are one of the most revolutionary innovations in blockchain technology, fundamentally changing how agreements and transactions are executed in a digital economy.

Defining Smart Contracts

At its core, a smart contract is a self-executing digital contract with the terms of the agreement directly written into code. These contracts run on blockchain networks, primarily Ethereum, and automatically execute when predefined conditions are met. Think of them as automated "if-then" statements embedded in a decentralized ledger. Once the conditions specified in the code are fulfilled, the contract executes itself without requiring any third-party intervention.

Smart contracts eliminate the need for intermediaries such as banks, brokers, or notaries, reducing costs, human error, and the risk of manipulation.

How Do Smart Contracts Work?

Smart contracts operate based on blockchain technology, ensuring transparency, security, and immutability. Here's a simple breakdown of how they work:

1. **Agreement Terms are Coded**: The terms and conditions of an agreement are written into a smart contract using a programming language, most commonly Solidity (on Ethereum).
2. **Deployment on the Blockchain**: The smart contract is deployed on the blockchain network, where it becomes immutable.

3. **Triggering Conditions**: The contract monitors events or conditions specified in the code.

4. **Automatic Execution**: Once the conditions are met (e.g., receiving payment, completing a task), the contract executes itself.

5. **Transaction Recorded**: The blockchain permanently records the transaction, creating a transparent and unchangeable history.

For example, imagine a freelance work agreement encoded into a smart contract: *"If the freelancer submits completed work and the client approves it, release payment to the freelancer."*

Who Can Use Smart Contracts?

In theory, anyone can use smart contracts, provided they have access to a blockchain network and understand the tools required to interact with smart contract platforms. However, deploying and managing smart contracts may require technical expertise in coding and blockchain architecture.

Platforms like Ethereum, Binance Smart Chain, and Solana have made it easier for developers to create smart contracts. Additionally, no-code or low-code platforms are emerging, enabling non-developers to interact with and even create basic smart contracts.

The Role of Solidity in Smart Contracts

Solidity is the most widely used programming language for creating smart contracts on the Ethereum blockchain. It is an object-oriented, high-level language influenced by Python, JavaScript, and C++.

Key Features of Solidity

- **Statically Typed**: Helps prevent runtime errors.

- **Supports Inheritance**: Code reuse across multiple contracts.
- **Libraries**: Shared code for common tasks.
- **Complex Data Structures**: Arrays, mappings, and structs.

Developers write Solidity code, which is then compiled into bytecode and executed on the Ethereum Virtual Machine (EVM).

For anyone looking to learn more about Solidity, the official documentation is an excellent resource: Solidity Documentation.

The Origins of Smart Contracts

The concept of smart contracts was first proposed by Nick Szabo in 1994. Szabo, a computer scientist and cryptographer, envisioned smart contracts as a way to digitally enforce agreements without intermediaries. He defined smart contracts as "computerized transaction protocols that execute the terms of a contract." In his vision, these protocols could handle synthetic assets, derivatives, and complex financial agreements. While Szabo's vision was ahead of its time, blockchain technology eventually provided the foundation to turn his ideas into reality, particularly with the launch of Ethereum in 2015.

Applications for Smart Contracts Across Industries

Smart contracts are transforming a wide range of industries, from finance to healthcare, real estate, and beyond. Below are some of the most prominent use cases:

1. **Finance and DeFi (Decentralized Finance)**: Smart contracts power decentralized exchanges (DEXs), lending platforms, and staking mechanisms, enabling peer-to-peer financial transactions without intermediaries.

2. **Supply Chain Management**: Smart contracts ensure transparency and accountability in supply chains. Each step, from manufacturing to delivery, can be tracked and verified on the blockchain.

3. **Real Estate**: Property sales, rental agreements, and escrow services can be managed through smart contracts, reducing paperwork and processing times.

4. **Healthcare**: Medical records can be securely stored and accessed using smart contracts, ensuring privacy while allowing authorized healthcare providers access.

5. **Voting Systems**: Smart contracts can make elections more transparent and secure by verifying voter identities and ensuring the integrity of votes.

6. **Intellectual Property and Royalties**: Artists, writers, and creators can automate royalty payments through smart contracts, ensuring they get paid every time their work is used or sold.

Advantages of Smart

Contracts • **Security**: Cryptographically secure, reducing

risks of fraud.

- **Transparency**: Transactions are publicly verifiable.
- **Efficiency**: Eliminates intermediaries and speeds up transactions.
- **Cost-Effective**: Reduces third-party fees.
- **Trustless**: Parties can transact without knowing or trusting each other.

Challenges and Limitations

While smart contracts offer immense potential, they are not without challenges:

- **Code Vulnerabilities**: Bugs or errors in the code can be exploited.
- **Scalability**: Blockchain networks can face congestion.
- **Legal Recognition**: Smart contracts may lack legal enforceability in some jurisdictions.

Smart contracts represent a fundamental shift in how agreements are executed and verified. They provide transparency, efficiency, and trust in a decentralized ecosystem. As blockchain technology continues to evolve, smart contracts will likely play an even more significant role in industries and day-to-day activities. Whether you're an investor, developer, or simply a curious reader, understanding smart contracts is essential for navigating the future of digital transactions. The era of automated, trustless agreements has arrived, and it's here to stay.

Real Estate and Smart Contracts

One powerful example of smart contract functionality is within the real estate industry. Imagine selling a house with a smart contract that automatically enforces a 10% royalty payment every time the property is resold. Traditionally, in real estate, sellers only make money from the initial sale of their property. However, with smart contracts, creators or original owners can continue to earn royalties on each subsequent transaction. This not only introduces a new revenue model but also creates an ongoing incentive for property owners to maintain or improve their assets.

Smart contracts eliminate intermediaries such as brokers, banks, or legal representatives, streamlining the sales process, reducing paperwork, and minimizing associated costs. Every transaction is securely recorded on the blockchain, ensuring transparency and immutability. Buyers and sellers can trust that the contract terms will be executed as coded, without the risk of fraud or manipulation.

Non-Fungible Tokens and Smart Contracts

Another industry experiencing rapid transformation due to smart contracts is the NFT (Non-Fungible Token) market. Non-fungible tokens (NFTs) are cryptographic assets built on blockchain technology, uniquely identifiable through distinct identification codes and metadata. Unlike traditional cryptocurrencies, which are fungible and can be exchanged one-to-one, NFTs are inherently unique and cannot be traded at equal value. This uniqueness makes NFTs ideal for representing digital art, music, collectibles, and even virtual real estate.

NFTs have gained immense popularity, representing digital ownership of art, music, videos, virtual real estate, and even intellectual property. With smart contracts, creators can embed royalty agreements directly into their NFTs, ensuring they receive a percentage of sales every time their digital assets change hands.

This royalty structure has revolutionized the creative industry, empowering artists and creators with financial control over their work. Previously, creators often lost out on revenue when their art or content was resold in secondary markets. Smart contracts ensure automated, transparent royalty payments without relying on intermediaries.

Governments and regulatory bodies are also paying close attention to the growth of NFTs and smart contracts. As these technologies continue to reshape industries, regulations are being developed to address legal, financial, and security concerns. Balancing innovation with compliance will be essential in ensuring smart contracts are widely adopted while maintaining consumer protection.

Celebrities, artists, musicians, and even everyday individuals like you and me can get involved in the NFT space and create substantial income opportunities. Later in this book, we'll share effective research methods and insider secrets to help guide you on this exciting journey.

Healthcare and Smart Contracts

Smart contracts have the potential to revolutionize several industries by enhancing transparency, efficiency, and security in operations. One of the key sectors benefiting from this innovation is healthcare. Blockchain technology allows encoded health records of patients to be securely stored with a private key, ensuring only authorized individuals can access sensitive data. This significantly improves privacy and security standards. Additionally, smart contracts enable automated sharing of hospital receipts with insurance companies as proof of service, streamlining the claims process. Beyond patient records, blockchain can be employed for managing medical supplies, tracking drug authenticity, and ensuring regulatory compliance, reducing inefficiencies and the potential for fraud.

Using Smart Contracts for Supply Chains

Supply chains are another area where smart contracts have made a significant impact. Traditional supply chain management often relies on paper-based systems, which require documents to pass through multiple intermediaries for approval. This outdated process is prone to delays, errors, and fraudulent activities. Recent global crises, such as the COVID-19 pandemic and the war in Ukraine, have exposed vulnerabilities in global supply chains. Blockchain technology addresses these issues by providing an accessible, transparent, and secure digital ledger. Smart contracts can automate tasks such as inventory tracking, payment processing, and shipment verification, ensuring that all stakeholders have real-time visibility into the supply chain.

Financial Services Use of Smart Contracts

The financial services industry is also experiencing a transformation driven by smart contracts. These digital agreements streamline and automate various financial processes, including insurance claims, trade clearing, and auditing. For instance, in the insurance sector, smart contracts perform automated error checks and initiate payments once claims meet predefined conditions, reducing delays and administrative overhead. In bookkeeping, blockchain technology prevents unauthorized changes to accounting records, ensuring transparency and accuracy. Furthermore, smart contracts enable shareholders to participate in corporate decision-making processes in a secure and transparent manner. Large financial institutions such as Bank of America, JP Morgan, Goldman Sachs, and even tech giants like Microsoft are actively exploring blockchain and smart contract solutions to optimize their operations.

However, there is an ongoing debate about centralized entities adopting blockchain technology. While their involvement can drive innovation and mainstream adoption, it also raises concerns about control and decentralization. The true power of blockchain lies in decentralized finance (DeFi), where individuals can freely transact using various digital currencies without relying on traditional banking intermediaries. The continued growth and adoption of DeFi solutions have the potential to challenge and reduce the dominance of centralized financial institutions.

Human Error in Smart Contracts

While smart contracts are designed to be secure and reliable, they are still created and programmed by humans. Any mistake in the code can result in costly consequences, as smart contracts are difficult to modify once deployed. Once a smart contract is finalized and added to the blockchain, it becomes almost impossible to alter. This means errors in the code can be both time-consuming and expensive to fix.

To mitigate these risks, it's essential to:

- Conduct thorough testing of your smart contract.
- Hire professional auditors who specialize in blockchain cybersecurity.
- Ensure multiple experts review the code before deployment.

Loopholes and Interpretation Challenges

Smart contracts are based on strict code execution, and the concept of "good faith" in traditional agreements doesn’t always translate seamlessly into digital contracts. Vague terms and unclear conditions can create loopholes, potentially allowing unethical exploitation of the contract.

To avoid this issue:

- Clearly define all terms and conditions.
- Avoid vague or overly complex clauses.
- Consult with legal professionals familiar with blockchain contracts.

Third-Party Involvement

One of the key promises of smart contracts is reducing the need for intermediaries. However, this does not entirely eliminate the need for third parties. Their roles have simply shifted. For example:

- Lawyers may not need to draft every individual contract but are still essential for reviewing legal implications of smart contract terms.
- Developers with expertise in Solidity are needed to translate contract terms into secure and functional code.

This redefined role means collaboration between legal experts and blockchain developers is crucial successfully implementing smart contracts.

Addressing Vague Terms

Traditional contracts often rely on subjective interpretations of terms and conditions. Smart contracts, however, require precise and unambiguous instructions. Any vagueness can lead to failures in execution.

To ensure clarity:

- Clearly articulate every condition in the contract.
- Use standard templates when possible.
- Make sure the purpose and utility of the contract are well-defined.

The Road Ahead

The rise of smart contracts marks the beginning of a new era of financial independence and digital empowerment. With proper education, due diligence, and a clear understanding of their capabilities and limitations, anyone can leverage smart contracts to build wealth, automate processes, and create long-term financial freedom.

So, what's stopping you? Start exploring the world of cryptocurrency and smart contracts today. Whether you're an artist, entrepreneur, or just someone looking for innovative investment opportunities, the tools for success are at your fingertips. Embrace this technology, chase your dreams, and build your path to financial freedom through blockchain and smart contracts!

Chapter 10

Staking, Lending and Borrowing:

Is Compound Interest The 8th Wonder of the World?

What if I told you that you could become your own bank today and beat the traditional banking system at its own game? Would you like to learn how to do that? In this chapter, we'll uncover how banks profit from your money and how, through the power of decentralized finance (DeFi), you can take control and replicate their strategies for your own financial freedom. Whether you're interacting with peer-to-peer (P2P) or engaging in business-to-business (B2B) transactions, DeFi empowers you to become the architect of your financial destiny.

How Banks Profit Off Your Money

Banks operate on a simple yet highly profitable model: they borrow money from depositors (YOU) and pay a modest interest rate on your savings. In turn, they lend out your deposited funds to borrowers at a significantly higher interest rate, profiting from the difference, known as the interest rate spread.

But it doesn't stop there. Banks also charge you monthly fees, overdraft fees, transaction fees, and many other hidden charges. These fees generate massive profits while offering you minimal returns on your deposits. Meanwhile, the interest they earn from borrowers isn't shared with you; instead, it's reinvested or lent out again to generate even more profit.

Diversification of Banking Services

Traditional banks don't rely solely on loans to make money. They also diversify their income streams by offering other financial services such as investment banking, wealth management, and financial advisory services. These services can generally be broken down into three key revenue streams:

1. **Interest Income:** Income made from interest is the primary source of revenue for most commercial banks. Banks use your deposits to provide loans and earn interest from borrowers. In exchange for keeping your money in their accounts, banks offer you a low interest rate, often not even enough to keep up with inflation.

Some banks, like credit unions, offer slightly better terms. Credit unions tend to have lower fees, better customer service, and higher interest rates on deposits because they are member-focused institutions rather than profit-driven corporations.

Interest income is influenced by supply and demand. When demand for long-term debt instruments (like loans or mortgages) is high, banks can offer lower interest rates to depositors while charging borrowers higher rates. When demand is low, they may increase the interest rates paid on deposits to attract more funds.

2. **Fees and Charges:** Banks earn billions annually through various fees, including account maintenance fees, overdraft charges, wire transfer fees, and ATM usage fees. These seemingly small fees accumulate into significant revenue streams for centralized banks.

3. **Inflation and Fixed-Term Deposits:** When inflation rises, banks benefit from locking in fixedterm deposits at low interest rates while lending out money at higher rates. If inflation decreases, they continue to profit because the rates they charge borrowers remain higher than what they pay depositors.

The DeFi Alternative: Be Your Own Bank

Now, imagine having the power to replicate these banking strategies without relying on intermediaries. DeFi allows you to do just that. Through decentralized platforms, you can stake your assets, provide liquidity, and earn interest rates that you choose and agree upon—all without needing a middleman or undergoing invasive Know Your Customer (KYC) protocols.

With DeFi:

- You control your assets: No bank can freeze or withhold your funds.
- You earn higher interest rates: By lending your assets directly to others, you eliminate intermediaries and maximize your returns.
- You avoid unnecessary fees: DeFi platforms often have lower transaction fees compared to traditional banking systems.

The Freedom of Staking

Staking in DeFi is like a savings account on steroids. You can lock your assets into a smart contract and earn Annual Percentage Rates (APR) that are significantly higher than what traditional banks offer. Best of all, you choose your terms, and no one can deny you access to these financial tools.

DeFi represents a revolution in finance. By understanding how banks profit from your money and leveraging decentralized financial tools, you can take control of your financial future. The question is no longer *"Can you beat the banks at their own game?"* but rather *"When will you start?"*

The tools are here. The opportunity is now. It's time to become your own bank and claim your financial freedom.

How Centralized Banks Profit from Their Operations

Before we dive deeper into staking and the power of compound interest, let's first address two other significant ways centralized banks make money from you. Albert Einstein once said, *"Compound interest is the eighth wonder of the world. He who understands it, earns it; he who doesn't, pays it."* This profound truth underscores the power of financial systems, but to truly understand how to benefit from it, we need to grasp how centralized banks profit from their operations.

Centralized banks often provide capital market services for financial corporations and investors. These capital markets serve as platforms where businesses seek funding to grow, expand, or finance largescale projects. In return, investors expect a return on investment (ROI) for the capital they provide. Centralized banks play a pivotal role in facilitating these activities through a range of specialized services, including Sales and Trading, Underwriting, and Mergers & Acquisitions (M&A) advisory services.

1. **Sales and Trading Services:** Centralized banks act as intermediaries in financial markets, buying and selling securities, stocks, and bonds on behalf of their clients. They profit from transaction fees, commissions, and the spread between buying and selling prices. These banks also often use their own capital to trade in financial markets, aiming to maximize returns.
2. **Underwriting Services:** In this role, banks help companies raise capital by issuing stocks or bonds. They assume the risk of buying the newly issued securities from the company and then reselling them to investors. The bank charges a fee for this service, and any difference between the purchase and resale price is additional profit.
3. **M&A Advisory Services:** Centralized banks advise corporations on mergers, acquisitions, and other strategic transactions. They guide businesses through the complexities of valuations, negotiations, and regulatory approvals, earning substantial fees for their expertise.

These services allow banks to leverage capital from depositors (like you and me) to generate substantial profits. While these financial operations are essential for global economic activity, they are also structured in a way that benefits the banks far more than their customers.

Now, imagine being able to participate in similar financial strategies through decentralized finance (DeFi). With DeFi, you have the ability to earn passive income, lock in interest rates, and grow your wealth without the restrictions and excessive fees imposed by traditional banking systems. No more KYC (Know Your Customer) roadblocks, no more denial of service based on arbitrary credit scores, and full control over your financial assets.

In the next section, we'll dive deeper into how staking works, how you can get started, and the potential it holds for building long-term financial freedom through decentralized finance. The future of banking is decentralized, and staking is your gateway to being your own bank.

What is Staking?

Staking is a method of earning passive income by participating in the verification of transactions on a blockchain network using specific DeFi cryptocurrencies. By holding and staking your digital assets, you contribute to the security and functionality of the blockchain, and in return, you earn rewards in the form of annual percentage yield (APY) or annual percentage rate (APR).

Staking yields significantly surpass what traditional savings accounts offer, making decentralized finance (DeFi) an attractive option for those seeking higher returns on their assets. In the world of cryptocurrency, staking allows users to earn passive income simply by holding and securing digital assets on a blockchain network. Here are some examples of popular cryptocurrencies and their potential annual percentage yields (APY):

- Solana (SOL): Earns approximately 8% per year.
- Ethereum (ETH): Offers around 5.5% yearly

- Cosmos (ATOM): Generates between 8-10% annually.
- Internet Computer (ICP): Provides returns of up to 15% annually.
- Polkadot (DOT): Offers returns of up to 14% annually.
- Algorand (ALGO): Earns an impressive 10% APY.
- Binance Coin (BNB): Can yield up to 27% per year.
- Luna (LUNA): Can yield anywhere from 5-25% APY.
- Tezos (XTZ): Typically earns between 5-6% annually.
- VeChain (VET): Offers returns ranging from 3-4%.
- Lisk (LSK): Generates yields of about 5-6% APY.
- Loom Network (LOOM): Also provides competitive staking rewards.

These staking rewards are possible because staked tokens play a critical role in maintaining and securing blockchain networks. By participating in staking, you essentially become part of the blockchain's consensus mechanism, such as Proof of Stake (PoS) or Delegated Proof of Stake (DPoS), which are used to validate transactions and secure the network.

Why Are Staking Yields So High Compared to Traditional Banks?

1. **No Intermediaries:** In DeFi, there are no middlemen like banks taking a large cut of the profits.

   Rewards are directly distributed to stakers.

2. **Blockchain Incentives**: Networks incentivize users to stake their tokens to ensure security and decentralization.

3. **Market Volatility**: High-risk environments often come with higher rewards, attracting users to participate.

4. **Dynamic Reward Structures**: Some blockchain projects offer higher yields during their early stages to encourage adoption and participation.

Staking isn't just about earning rewards; it also empowers individuals to actively participate in governing blockchain networks. By staking tokens, users often gain voting rights on proposals, upgrades, and other governance-related decisions.

Risks and Considerations in Staking

While the rewards are appealing, it's essential to understand the potential risks:

- **Market Volatility**: Cryptocurrency prices can be highly volatile.
- **Lock-up Periods**: Some staking platforms require users to lock their assets for a certain duration.
- **Technical Risks**: Errors in smart contracts or network failures can lead to asset loss.

However, with careful research and diversification across multiple staking opportunities, the benefits can far outweigh the risks.

The financial freedom offered by decentralized finance (DeFi) is closer than ever, and staking is one of the most powerful tools to achieve it. On some platforms, staking yields can reach up to 25% APY Cryptocurrencies like Decred offer returns of 7-10% APY, and ICON can yield between 6-36% annually. These are just a few examples of hundreds of cryptocurrencies that allow you to earn passive income through staking. Compared to the meager returns offered by traditional banks, these APYs make a compelling case for moving your funds into staking platforms and becoming your own bank.

Here are some of the top staking tokens by market capitalization:

- Cardano
- Near Protocol
- SUI
- CRO
- Cosmos
- Injective
- Serum
- Etheruem
- Solona
- USDT
- Avalanche
- Polkadot

These tokens represent only a fraction of the opportunities available for earning passive income through staking.

Benefits of Staking

- **Earn Passive Income**: An easy way to earn interest on your cryptocurrency holdings.
- **No Equipment Needed**: Unlike crypto mining, staking doesn't require expensive hardware.
- **Support Blockchain Security**: Your staking helps secure and maintain the blockchain network.
- **Environmentally Friendly**: Staking consumes far less energy compared to mining.

Flexible Terms: Choose how long you want to stake your funds, with the option to withdraw at any time.

- **Financial Autonomy**: Become your own bank and engage in peer-to-peer lending and borrowing.

Staking empowers users by offering more benefits to them than to the service providers. To help you get started, here are some of the top crypto staking platforms.

Top Crypto Staking Platforms

- Crypto.com: Best platform for flexible withdrawals.
- Aqru: User-friendly interface for easy staking.
- eToro: Regulated platform with low fees.
- Coinbase: Trusted exchange with staking services.
- Binance: High staking rewards.
- Nexo: Earn up to 8.5% APY on holdings.
- Kraken: Attractive on-chain staking yields.
- Gemini: Trusted for trading and staking.
- Ledger: Best for cold storage staking.
- Ellipal: Secure cold storage wallet.
- Exodus: Multi-currency wallet with desktop and mobile support. Atomic Wallet: Supports staking for Tezos, Cosmos, NEO, and more.
- Trust Wallet: Mobile wallet supporting multiple staking tokens.
- Lido: Best Ethereum liquid staking platform

- Aave: Best decentralized lending and staking protocol
- Rocket Pool: Best for Ethereum node staking
- KuCoin: Best for altcoin staking
- MEXC: Best for staking newly listed coins
- Bybit: Best for multiple crypto earn products
- Keynode: Best liquid staking platform with attractive welcome bonus ($100)

While many of these platforms are reputable, if security is your top priority, Ledger and Ellipal are the most logical choices. These hardware wallets store your funds offline in cold storage, offering maximum protection against hacks and cyber theft.

Start staking today, take control of your financial future, and let your money work for you instead of sitting idle in traditional bank accounts.

Crypto Lending

So, what other things can you do now that you are your own bank? Imagine you could lend your money to other investors like the banks do with your money, but instead, now you're the one earning interest. Well folks, I'm here to share with you that you can do that, and you can actually get started today.

Crypto Lending is a type of decentralized finance (DeFi) that allows you to lend your cryptocurrencies to different borrowers. This allows you to earn interest payments in exchange, also called crypto dividends. Many different platforms specialize in lending crypto, also offering stablecoins that earn up to 19% APY for using them, on top of various other cryptocurrencies to choose from.

Let's go through an example of how this would work. Let's say you have 40 Bitcoin. You want to make some steady gains from it, so you decide to deposit it into a credible crypto lending exchange. As such, you find an investor you want to lend to, and you choose how long you want to lock them in for—every month, week, or even daily—depending on the platform and terms.

Those platforms will tell you how long funds have to be locked in at a rate. As long as they agree to the terms and the lock in time and rate, you will receive interest, just like how the banks would operate when you deposit funds with them, and they lend it at higher interest rates to others.

It is important again to mention that these interest rates can differ depending on the platform, typically ranging from 3-8%, and in some cases, can go quite higher, up to 19% APY—far exceeding what centralized banks can offer.

A benefit to crypto lending is that borrowers can also have the chance to stake their cryptocurrency as guarantees of loan repayment or as collateral. Thus, the investors will be able to sell the crypto assets in case the borrower doesn't pay off the loan, enabling them to recover their losses. Platforms have shown lenders that they have a high chance of receiving their funds back in such cases because they ask borrowers to stake 25-50% of the loan in cryptocurrencies. This provides confidence to those who loan crypto to borrowers who might have a high risk of defaulting.

How Does Crypto Lending Work?

Crypto lending works through decentralized finance (DeFi) or DApps (Decentralized Applications) via a third party that connects lenders and borrowers.

1. **The Lenders**: Represent the first party involved in the lending of cryptocurrencies. They might be investors who want to grow their assets and so they lend their holdings out to earn interest.

2. **The Crypto Lending Platform**: This represents the second party, where lending and borrowing transactions are facilitated and processed.

3. **The Borrowers**: Represent the third party in the process, and they are the ones who will receive the funds. They could either be businesses that need funding or just a normal person looking for financial support.

The Crypto Lending Process

1. **Deposit Funds**: The lender deposits their crypto assets into a lending platform.

2. **Loan Agreement**: Borrowers apply for loans, often with collateral in cryptocurrency.

3. **Terms Agreement**: Both parties agree on terms such as interest rates, loan period, and collateral.

4. **Funds Transfer**: The platform facilitates the transfer of funds to the borrower.

5. **Interest Payments**: Borrowers make interest payments based on the agreed-upon schedule.

6. **Loan Repayment**: At the end of the term, the borrower repays the loan, including any accrued interest.

7. **Collateral Liquidation** (if necessary): If the borrower defaults, the collateral is liquidated to repay the lender.

By participating in crypto lending, you are not only earning passive income but also contributing to the growth and stability of the decentralized financial ecosystem. With the right research, diligence, and selection of credible platforms, you can start building financial freedom today through crypto lending.

More On Crypto Lending

- The borrower goes to a decentralized exchange or dApp and requests a crypto loan.
- The borrower stakes the crypto collateral for the loan as soon as the loan request is accepted by the platform. Until they are able to repay the entire loan, the borrower will not have the opportunity to retrieve their collateral.
- Using the platform, the lenders will automatically fund the loan, which is a process that investors typically cannot see.
- Investors will receive regular interest payments, similar to how traditional banks earn from lending funds.
- When the borrower manages to pay off the whole loan, they will get back their crypto collateral.

Every DeFi exchange offers its own types of lending cryptos, terms, and different APR rates, just like traditional banks offer varying interest rates. Remember, since different crypto lending exchanges have varying rates, it's ideal to find the ones that can yield the best return on your investment. The platform you choose will directly affect your ROI, and each comes with its own risks.

It's crucial to DYOR (Do Your Own Research) and consider using multiple platforms to spread risk. Diversifying your portfolio and investments across several platforms can safeguard your returns and reduce potential losses.

Typically, annual yields from crypto lending can range from 3% to 8% for cryptocurrencies, while stablecoins often offer higher returns, varying from 10% to 19%. Different platforms offer different rates depending on the coin being lent, so optimizing your returns will require selecting the right platform based on your holdings.

Here is a list of some trusted Crypto Lending exchanges you can use today to start earning interest:

- Nexo- Best for users concerned about security
- Binance- Best for using multiple Cryptocurrencies
- Coinloan- Best for users looking for Trust and Transparency
- Aave- Best for Decentralized Lending
- Compound- World of crypto protocols for lending and borrowing
- MakerDao- 25+ crypto assets as collateral
- YouHolder- 90% loan-to-value ratio (LTV)
- MoneyToken- Manage all your crypto assets and also receive cryptocurrencies loans with a few clicks

These crypto lending exchanges are quite trustworthy, and you can get started today to become your own bank, beating the traditional banks at their own game while earning financial freedom for yourself and your family.

Leveraging Your Assets Through Borrowing

The last strategy I want to discuss is leveraging your assets through borrowing. This process involves using cryptocurrency as collateral, and it's crucial to select the right platform. While many platforms offer crypto borrowing services, not all of them are trustworthy. We've already listed several credible options above to help you get started. However, it's essential to *Do Your Own Research (DYOR)* to ensure the platform is safe, reliable, and legitimate before proceeding.

Once you've identified a credible decentralized finance exchange, the next step is to verify whether the platform supports borrowing the specific cryptocurrency you're interested in. Not every platform offers all cryptocurrencies, and returns can vary significantly depending on the coin and platform.

Taking out a crypto loan is generally much easier compared to traditional loans. The amount you can borrow depends on the value of the collateral you provide. This relationship is referred to as the *loan-tovalue (LTV) ratio*. For example, if you deposit $20,000 worth of cryptocurrency as collateral and receive a $10,000 loan, the LTV ratio is 50%.

Crypto loans typically come with relatively low LTV ratios due to the inherent volatility of the crypto markets. This protects the lender by ensuring they can recover their funds if the borrower defaults, as they can liquidate the collateral.

By leveraging these borrowing and lending strategies, you can start compounding your assets and paving the way towards financial freedom. You no longer need to rely on traditional banking services—you can now control your financial future, build wealth, and achieve financial confidence through decentralized finance.

Chapter 11

Stablecoins

What if I could show you how to keep your funds relatively stable when cryptocurrency markets turn bearish or start trading sideways? Would you like to take advantage of that? In this chapter, I'll discuss the power of stablecoins and how they can help you maintain financial stability during market corrections.

A Brief History of Stablecoins

The first stablecoin, Tether (USDT), was created in 2014, and many subsequent stablecoins have followed its model. Tether (USDT) is a blockchain cryptocurrency whose tokens in circulation are backed by an equivalent amount of U.S. dollars, maintaining a price peg of 1 USDT = 1 USD. Stablecoins track traditional fiat currencies, such as the dollar, euro, and Japanese yen, which are held in designated bank accounts.

Interestingly, national banks are now allowed to participate in public decentralized networks and use stablecoins for payment settlements. According to guidance from the Office of the Comptroller of the Currency (OCC), national banks and federal savings associations may use blockchain technologies and stablecoins to perform permissible banking functions. This means your current bank might already be leveraging stablecoins for their internal operations to stabilize trade accounts during market volatility.

What Are Stablecoins?

Stablecoins are a type of cryptocurrency designed to offer price stability. While other cryptocurrencies can experience extreme volatility, stablecoins remain pegged to an underlying asset, typically at $1.00 per token.

Some stablecoins are backed by reserves of the asset they represent, while others use algorithms or alternative mechanisms to maintain their peg.

Let's consider an example:

- Mike has $100,000 invested in cryptocurrencies such as Bitcoin, Ethereum, and Solana.
- He made substantial profits during a bullish cycle but anticipates a market correction.
- To protect his gains, Mike converts his entire portfolio into a stablecoin like Tether (USDT).
- Now, his $100,000 remains secure and unaffected by market downturns.
- When the market dips, Mike can use his stablecoins to buy assets at lower prices, repeating this cycle to build long-term financial freedom.

How Do Stablecoins Work?

Stablecoins maintain their value through various mechanisms and collateral types. Depending on their backing, they can carry different risk levels:

1. **Fiat-Backed Stablecoins**: Backed by traditional currencies like the U.S. dollar, these stablecoins are considered the most stable because they are tied to centralized financial systems.

2. **Crypto-Backed Stablecoins**: Collateralized by other cryptocurrencies, these stablecoins can be more volatile due to fluctuations in their underlying assets.
3. **Precious Metal-Backed Stablecoins**: Pegged to assets like gold or silver, these stablecoins offer an alternative hedge against inflation.
4. **Algorithmic Stablecoins**: These rely on algorithms and smart contracts to control supply and demand, maintaining their peg without physical collateral.

Understanding these categories will help you make informed decisions when integrating stablecoins into your portfolio. Stablecoins are not just a safe haven during bearish markets—they are also powerful tools for strategic financial management in decentralized finance.

Fiat-Backed Stablecoins

Fiat-backed stablecoins use your dollars (or other fiat currency) to buy stablecoins that you can redeem later for your original currency. Fiat-backed stablecoins maintain a fiat currency reserve, like the U.S. dollar, as collateral to issue a suitable number of crypto coins.

Other forms of collateral, which we will cover, can include precious metals like gold or silver, as well as commodities like oil, although most of the present-day fiat-collateralized stablecoins use dollar reserves. A few fiat stablecoins to mention are Tether (USDT), USD Coin (USDC), and Binance USD (BUSD). Unlike other cryptos, whose value can fluctuate widely, fiat-backed stablecoins aim to have very small price fluctuations. Not all fiat stablecoins are trustworthy; it is important that you DYOR (Do Your Own Research) and look for ones that have a proven track record.

Crypto-Backed Stablecoins

Another form of stablecoins is crypto-backed stablecoins. These are backed by crypto assets. A few examples of crypto-backed stablecoins are DAI, TerraUSD, and TrueUSD. Since they are backed by crypto assets, they can tend to be more volatile.

Crypto-backed stablecoins are often over-collateralized to ensure the stablecoin's value. For example, $200 worth of a crypto-backed stablecoin may be tied to an underlying crypto asset worth $300, so if the underlying crypto loses value, the stablecoin has a built-in cushion and can remain at $200. These assets are less stable than fiat-backed stablecoins, and it's a good idea to always keep track of the underlying crypto asset behind your stablecoin and how it is performing in the marketplace.

Precious Metal-Backed Stablecoins

Precious metal-backed stablecoins are a groundbreaking innovation in the world of digital assets, offering investors a convenient way to gain exposure to precious metals, particularly gold, without the associated complications of physical storage and transportation. These stablecoins are backed by the value of gold, allowing investors to enjoy the benefits of owning precious metals without the hassle of dealing with the logistics of securing, storing, and insuring the metal.

- **No Storage or Transportation Hassles**: Investors can gain exposure to the price movements of gold without worrying about safely storing or transporting the physical metal.
- **Accessibility and Liquidity**: Precious metal-backed stablecoins allow you to buy, sell, and transfer your gold holdings easily and instantly, providing liquidity that physical gold cannot offer.
- **Protection Against Inflation**: Gold has long been viewed as a hedge against inflation. By using gold-backed stablecoins, investors can maintain their purchasing power while reducing the risk of volatility.

- **Easier to Divest and Diversify**: These tokens allow investors to diversify their portfolios with an asset that typically has low correlation with traditional financial markets, such as stocks and bonds.

By adopting these digital representations of precious metals, such as gold-backed stablecoins, investors can access the value of gold without the high costs and risks associated with physical gold ownership.

Here is a list of current precious metal-backed stablecoins:

- Tether Gold (XAUT): 1:1 troy fine ounce of gold. Current price when written: $1,946.
- DigixGlobal (DGX): 1:1 gram of gold. Current price per token when written: $31.44.
- Paxos Gold (PAXG): 1:1 fine troy ounce of gold. Current price when written: $1,951.76.
- GoldCoin (GLC): 1:1000 GoldCoin per ounce of gold. Current price per token when written: $0.12.
- The Perth Mint Gold (PMGT): 1:1 oz GoldPass certificate. Current price per token: $1,997.

Algorithmic Stablecoins

Algorithmic Stablecoins represent a different approach to maintaining price stability. Unlike precious metal-backed stablecoins, algorithmic stablecoins are not tied to any physical asset. Instead, they use smart contracts and algorithms to adjust the supply and demand of tokens in real-time to keep their value stable, usually pegged to a specific value, such as $1 USD.

- **How They Work**: These stablecoins adjust their circulating supply by creating or burning tokens based on price fluctuations. If the price goes above the peg (for example, above $1), the algorithm increases the supply of tokens, bringing the price down. Conversely, if the price falls below the peg, the algorithm reduces the supply to drive the price back up.

- **Decentralization**: Algorithmic stablecoins, like their counterparts in the crypto world, are decentralized and operate independently of traditional financial systems. This means they are not subject to the same regulatory rules or central government control, which appeals to those seeking privacy, autonomy, and access to decentralized finance (DeFi) platforms.
- **Market Stability**: While it may seem counterintuitive, these stablecoins have the potential to withstand market volatility in ways that other cryptocurrencies cannot. By automating adjustments to supply, they help stabilize their value during market shocks, offering investors a potentially more reliable store of value during turbulent times.
- **Accessibility and Growth**: Algorithmic stablecoins provide an opportunity to participate in DeFi applications without needing a bank account or traditional financial services. Investors can engage in staking, lending, and earning annual percentage yields (APY) without facing the typical barriers, such as Know Your Customer (KYC) processes, credit checks, or geographical restrictions.

Examples of algorithmic stablecoins include AMPL (Ampleforth), MakerDAO, FEI, RAI, and FRAX, all of which have demonstrated the ability to handle market shocks and fluctuations in demand effectively.

Key Benefits of Algorithmic Stablecoins

1. **Volatility Resistance**: Algorithmic stablecoins are designed to handle price fluctuations and can remain stable during bearish market trends.

2. **Enhanced Liquidity and Mobility**: Unlike traditional assets, algorithmic stablecoins are easily transferable, offering a high degree of mobility in the digital economy.

3. **Staking and Passive Income**: Staking these stablecoins can generate rewards, with annual yields reaching up to 19% depending on the platform and token. Staking essentially allows holders to lend out their tokens to support the network and earn passive rewards.

4. **No Traditional Financial Barriers**: These stablecoins offer access to financial services without requiring a credit history, KYC, or other traditional banking processes.

5. **Private and Secure**: The decentralized nature of algorithmic stablecoins ensures privacy and avoids the need for centralized financial services, reducing the risk of data exposure or high fees.

How to Get Started with Stablecoins

To get started with precious metal-backed or algorithmic stablecoins, you need access to a crypto exchange or a digital wallet that supports stablecoins. Some centralized exchanges like Coinbase, SoFi, Gemini, Crypto.com, Nexo, and Binance offer fiat-backed stablecoins, while decentralized exchanges (DEXs) may provide a wider selection of stablecoins, including those that are algorithmic or precious metalbacked.

You can also swap tokens directly on decentralized platforms or use hardware wallets to manage your stablecoin holdings securely. It's important to check whether certain services are available in your location, as some exchanges may have geographical restrictions.

A Diverse Selection of Stablecoins

As the cryptocurrency landscape continues to evolve, stablecoins have emerged as a crucial part of the ecosystem, offering investors a way to participate in digital assets while maintaining a more predictable value. The diversity of stablecoins available today allows investors to select tokens that best align with their preferences and investment strategies. Some of the top stablecoins by market capitalization include:

- Tether (USDT)
- USD Coin (USDC)
- Binance USD (BUSD)
- TerraUSD (UST)
- Dai (DAI)
- TrueUSD (TUSD)
- Pax Dollar (USDP)
- Neutrino USD (USDN)
- Fei USD (FEI)
- Gemini Dollar (GUSD)
- Tribe (TRIBE)
- Frax (FRAX)
- Liquidity USD (LUSD)
- HUSD (HUSD)
- Origin Dollar (OUSD)
- Stasis Euro (EURS)

- sUSD (SUSD)
- Celo Dollar (cUSD)
- USDK (USDK)
- Magic Internet Money (MIM). This wide variety of stablecoins offers investors many options, whether they seek a USD-pegged stablecoin or those pegged to other currencies, such as the Euro (EURS). Additionally, stablecoins are not just a tool for stability—they also offer the opportunity for earning additional APY through staking. This ability to earn passive income by holding and staking stablecoins is a key benefit, as it provides the opportunity to receive rewards while maintaining a relatively stable value.

The Growth and Scrutiny of Stablecoins

The stablecoin market has grown rapidly, with a current market cap of around $184 billion (at the time of writing). As stablecoins continue to grow in both size and influence, they have come under increasing scrutiny from regulators. Given their substantial market cap and potential to disrupt traditional financial systems, regulators are starting to take a closer look at these digital assets.

In 2021, the International Organization of Securities Commissions (IOSCO) recommended that stablecoins be regulated as financial market infrastructure, similar to payment systems and clearinghouses. This is due to the potential systemic risks posed by stablecoins that could disrupt payment systems and settlement transactions if not properly regulated. This growing regulatory attention highlights the need for stablecoins to be managed and controlled responsibly, especially those deemed systemically important.

Why Stablecoins Empower Financial Freedom

Despite the regulatory concerns, stablecoins offer a unique opportunity for individuals to take control of their financial future. One of the key advantages is that stablecoins allow you to become your own bank, bypassing traditional financial institutions. Banks typically offer minimal interest on savings accounts, but with stablecoins, you can stake your holdings and earn annual percentage yields (APY), often much higher than traditional bank interest rates.

This ability to stake stablecoins and earn interest is a game-changer. By staking, you essentially provide liquidity or participate in the network, helping to secure the blockchain while earning passive rewards. The best part? You get to keep that interest, which is something traditional banking institutions often don't offer. With stablecoins, you can replicate the very processes that banks use to profit from your money, putting financial control back in your hands and helping you achieve financial freedom.

The Importance of Due Diligence (DYOR)

While the benefits of stablecoins are clear, it's crucial to do your own research (DYOR) before investing. Not all stablecoins are created equal, and it's essential to choose ones with strong reputations and credibility. Some stablecoins may have higher risks than others, so finding trusted and credible stablecoins (like the ones mentioned above) will help ensure you're making a safe investment.

By carefully selecting stablecoins with solid backing and operational mechanisms, you can begin taking control of your financial destiny. With the ability to earn APY and the security of a relatively stable asset, stablecoins offer a powerful tool for those seeking to build wealth, diversify their portfolios, and access the decentralized financial ecosystem.

Stablecoins are a compelling option for diversifying investments, earning passive income, and accessing decentralized financial services. As the stablecoin market continues to grow, investors have the opportunity to benefit from the stability of these digital assets while avoiding the traditional banking system's limitations.

By staking stablecoins, you can generate returns that traditional savings accounts can't match, all while keeping your money in a secure and decentralized environment. However, it's crucial to approach these opportunities with caution—always perform thorough research before committing to any stablecoin investment. Start taking control of your money, earning rewards, and building your financial future with stablecoins today!

Chapter 12

NFTs: Turning JPEGs into Millions

The Rise of NFTs and How to Capitalize on This Trend

You've probably heard the buzz about people making millions of dollars selling digital art as NFTs

on platforms like OpenSea. Believe it or not, these stories are real. In fact, the NFT market has experienced a massive surge in the last few years, reaching impressive figures. According to The Guardian, in 2021, collectors and traders spent $22 billion on NFTs, up from just $100 million in 2020. This explosive growth, as reported by DappRadar, underscores the meteoric rise of NFTs and their potential as a financial opportunity. In this chapter, we'll explore what NFTs are, the history behind them, where to conduct proper research, and how you can start making a profit from this booming trend in the crypto space.

What Are NFTs?

NFT stands for Non-Fungible Token, a digital asset that represents ownership or proof of authenticity of a unique item or piece of content. Unlike cryptocurrencies such as Bitcoin or Ethereum, which are interchangeable and fungible, each NFT is one-of-a-kind or part of a limited edition. This uniqueness is what gives NFTs value, especially when they represent digital art, collectibles, or even realworld assets like property or tickets.

In the NFT space, people can buy, sell, and trade these digital assets. These tokens live on a blockchain, typically Ethereum, ensuring that ownership is transparent and secure. NFTs can represent:

- Digital Art
- Collectibles (e.g., trading cards, virtual goods)

- Virtual Real Estate (e.g., Decentraland, The Sandbox)
- Music and Videos
- Domain Names
- Event Tickets
- In-Game Assets

The sheer variety of assets being tokenized into NFTs is a testament to the flexibility of the technology.

The History of NFTs: From Colored Coins to Today

The concept of NFTs has roots that go back nearly a decade. In 2012, the first NFT-like concept emerged as Colored Coins. These were small denominations of Bitcoin, as small as a single satoshi (the smallest unit of Bitcoin), which could represent various assets and have multiple use cases, such as: Property Rights

- Coupons
- Cryptocurrency Creation
- Company Shares
- Digital Collectibles
- Access Tokens
- Ownership and Transfer of Ownership

The idea of Colored Coins was initially explored in a blog post by Yoni Assia, and later, in 2013 and 2014, researchers like Meni Rosenfeld further elaborated on the potential of this new asset class. In

2014, as the concept began to develop, Kevin McCoy, a digital artist, minted the first known NFT, Quantum, a pixelated octagon that pulsed with shapes. This piece of art went on to sell for over $1.4 million at a Sotheby's auction in 2021.

Simultaneously, other innovators were building platforms that would allow the creation of digital assets. Counterparty, founded by Robert Dermody, Adam Krellenstein, and Evan Wagner in 2014, enabled asset creation on top of the Bitcoin blockchain. It featured a decentralized exchange and crypto tokens (XCP) and allowed users to mint their own NFTs.

Key Milestones in the Evolution of NFTs

The development of NFTs was an ongoing process throughout the 2010s. A few important milestones include:

- 2014: Kevin McCoy's "Quantum" becomes the first known NFT, marking a historic moment for digital ownership.
- 2015: Spells of Genesis, a blockchain-based trading card game, became one of the first NFTs on the Counterparty platform. The game launched the BitCrystals token as in-game currency, integrating NFTs into the gaming world.
- 2016: Counterparty teamed up with Force of Will, a popular trading card game, to issue digital cards as NFTs. The partnership highlighted the increasing potential for mainstream adoption of NFTs.

These early efforts laid the foundation for the growing interest in NFTs, particularly as digital collectibles gained traction.

How to Profit from NFTs Today

If you want to take advantage of the NFT boom, here's how you can get started:

1. **Research**: Before diving into the NFT space, do thorough research. Explore the different types of NFTs available, check out NFT marketplaces like Opensea, Rarible, and SuperRare, and track the performance of various collections. Be sure to study the artist or project behind each NFT and understand the potential for long-term value.

2. **Create or Collect**: Depending on your interests, you can either create your own NFTs (if you're an artist or content creator) or purchase NFTs to build a collection. Many people have made significant profits by flipping NFTs—buying them at a low price and selling them when demand increases.

3. **Staking and Earning Passive Income**: Some NFT platforms allow you to stake your NFTs or earn passive income through different mechanisms, such as lending your NFTs out or using them in games and virtual worlds.

4. **Be Aware of Scams**: Like any emerging market, the NFT space is rife with potential risks. Be cautious and avoid purchasing from unverified sources or participating in projects with little backing. DYOR (Do Your Own Research) to ensure that you're investing in legitimate projects.

Platforms for NFT Research

To properly research and track NFTs, use platforms like:

- **DappRadar:** Provides regular reports on NFT sales and has a leaderboard of the top NFT sales.

- **NFTStats**: A useful tool for tracking the market performance of NFTs.

- **OpenSea**: The largest NFT marketplace, offering insights into top-selling NFTs and emerging trends.
- **Rarible**: Another popular platform to buy, sell, and mint NFTs.
- **SuperRare**: A curated marketplace for high-end digital art NFTs.

NFTs are an exciting new frontier in the world of digital assets, and their potential for both creators and investors is vast. With the NFT market already surpassing $22 billion in 2021, there is no doubt that the industry will continue to grow. The history of NFTs, from Colored Coins to the first pixelated art piece sold for millions, highlights how far the technology has come in just a few years.

Whether you're an artist looking to monetize your work, a collector hoping to acquire valuable digital assets, or someone simply trying to profit from the NFT trend, research and awareness are key. By understanding the space, learning how NFTs work, and staying informed about market trends, you can make educated decisions and take part in the exciting world of NFTs.

The Evolution of NFTs: From Memes to a Multi-Billion Dollar Industry

The NFT (Non-Fungible Token) space has seen explosive growth over the past decade, with several key moments that shaped its evolution and widespread adoption. The journey started with simple ideas and niche projects, but today, NFTs have become a major force in the blockchain and cryptocurrency ecosystems. In this chapter, we'll take a closer look at the pivotal events in the history of NFTs, how meme culture contributed to their rise, and how Ethereum's introduction helped establish NFTs as we know them today.

The Birth of Meme NFTs and the Rare Pepe Phenomenon

The concept of NFTs began to gain significant traction in 2016, when people started issuing "Rare Pepes" on the Counterparty platform. Rare Pepes are memes featuring a character of a frog named Pepe, which became an internet sensation. These unique, collectible meme cards were turned into NFTs, and they sparked an entire subculture within the crypto and meme communities. As the popularity of Rare Pepes grew, so did the idea of creating more valuable and unique digital assets based on memes.

The Rare Pepe Meme Directory is an example of how the meme culture within the NFT space has evolved. It's a marketplace dedicated to trading these digital collectibles and showcases how memes can transcend their casual, humorous origins to become valuable digital assets. Counterparty, the platform where Rare Pepes were first issued, still supports numerous projects built on its infrastructure, many of which revolve around NFT assets.

Ethereum and the Emergence of ERC Standards

In 2015, the launch of the Ethereum blockchain marked a significant turning point for NFTs. Ethereum introduced the idea of smart contracts, which allowed developers to create custom tokens with unique properties and use cases. The Ethereum blockchain also introduced new token standards, such as ERC-721 and later ERC-1155, which enabled the creation of NFTs.

The ERC-721 token standard became the foundation of most NFTs. It defines the framework for creating non-fungible tokens, meaning each token is unique and cannot be exchanged for another one of equal value. These tokens are stored on the blockchain, making them immutable and traceable, which gives them inherent value in the world of digital ownership.

Later, ERC-1155 was introduced as an improvement to ERC-721, allowing for the creation of semi-fungible tokens that could represent both fungible and non-fungible assets within the same contract. This opened up even more possibilities for developers and collectors.

The Rise of Meme-Centric Projects in 2017

The year 2017 was a pivotal year for the NFT market, with several significant projects and releases

that helped propel NFTs into the mainstream. One of the most notable was the launch of Curio Cards in May 2017, one of the first NFT projects to use the Ethereum blockchain. Curio Cards were digital art collectibles, and they marked the beginning of the widespread adoption of Ethereum-based NFTs.

In June 2017, CryptoPunks, created by Larva Labs (John Watkinson and Matt Hall), debuted. CryptoPunks are a collection of 10,000 unique 24x24 pixelated characters that quickly became one of the most iconic NFT collections in history. CryptoPunks continue to command high prices today and are a symbol of the NFT revolution.

Another major milestone in 2017 was the launch of CryptoKitties by Dapper Labs in November. CryptoKitties allowed users to collect, breed, and trade virtual cats. It was one of the first successful NFTbased games and became a viral sensation. CryptoKitties' popularity helped raise awareness about NFTs and their potential as digital collectibles.

In August 2017, Mooncats debuted as the first-ever NFT cats, introducing bonding curves and naming mechanisms for the first time. Decentraland, a virtual world built on Ethereum, also launched its land token sales in August 2017, paving the way for the integration of NFTs into virtual worlds. These early projects showed how NFTs could extend beyond simple art collectibles to virtual goods, gaming, and even real estate.

The Launch of OpenSea and the NFT Marketplace Boom

By December 2017, OpenSea was launched, quickly becoming one of the first and most popular decentralized exchanges and marketplaces for digital collectibles. OpenSea allowed users to buy, sell, and trade NFTs on a decentralized platform, helping to kickstart the NFT ecosystem. The launch of OpenSea was a game-changer, as it provided a single platform for people to explore and purchase a wide range of NFTs, from art and music to virtual goods and gaming assets.

The Explosive Growth of NFTs in 2018 and 2019

The NFT space saw dramatic growth in 2018 and 2019 as more developers and artists entered the scene. 200+ projects had emerged by the end of 2019, with many focusing on creating new and innovative NFT concepts. The rise of marketplaces like OpenSea, Rarible, and Foundation enabled artists and creators to directly monetize their work through the sale of NFTs, creating a thriving ecosystem for digital creators and collectors.

The increasing popularity of digital collectibles and the integration of NFTs into games, virtual worlds, and art allowed the ecosystem to continue evolving. Projects began to explore new use cases for NFTs, including music albums, virtual real estate, and even celebrity-backed NFTs.

Looking Ahead: The Future of NFTs

Today, NFTs continue to evolve, with new projects and applications emerging regularly. The 200+ projects in the space in 2019 have now expanded even more, and the number of NFT marketplaces, platforms, and collectors has skyrocketed. NFTs are no longer just a niche market for digital art or meme enthusiasts—they've become a mainstream industry that spans gaming, entertainment, fashion, sports, and much more.

The rise of NFTs has also prompted other blockchain platforms, such as Solana, Binance Smart Chain, and Flow, to create their own ecosystems for NFTs, diversifying the landscape even further.

Embrace the NFT Revolution

The history of NFTs is a testament to the power of innovation, community, and the growing influence of blockchain technology. From the early days of Rare Pepes on Counterparty to the launch of OpenSea and the rise of major NFT projects like CryptoPunks, CryptoKitties, and Curio Cards, the evolution of NFTs has been nothing short of revolutionary.

As the NFT ecosystem continues to expand and mature, opportunities for both creators and collectors will continue to grow. Whether you're an artist looking to tokenize your work, a gamer seeking virtual assets, or an investor hoping to profit from the NFT boom, now is the time to get involved in this exciting and rapidly evolving space. The future of NFTs is bright, and the possibilities are endless.

The Evolution of NFTs and Play-to-Earn Games in the Metaverse

The journey of NFTs has experienced exponential growth since their humble beginnings. As we moved into 2019, collections like Cryptoskulls started to thrive, and within just a few years, they gained significant traction. In January 2022, Cryptoskulls broke records by surpassing OpenSea's 24-hour sales volume chart, achieving 5,000 ETH in sales, equating to approximately $16 million.

By then, Cryptoskulls had reached 6,800 ETH in trading volume, with a floor price of 0.74 ETH ($2,813) per piece. Although the trade volumes of NFTs are still relatively small compared to other crypto markets, the NFT market has shown remarkable growth, and its momentum is only increasing.

As NFT onboarding becomes easier, the barrier to entry has been lowered significantly. Web3 wallets, like MetaMask, are continuously improving, making it easier for newcomers to interact with NFTs and blockchain networks. Additionally, Dapper Labs introduced the Dapper wallet, which has eliminated the need for gas fees, simplifying the transaction process for users.

On top of that, numerous NFT resource websites like nonfungible.com, nftcryptonews.com, and icy.tools provide essential tools, market metrics, gameplay guides, and general information, giving users the knowledge to make informed decisions in the space.

The Rise of Virtual Worlds and Play-to-Earn (P2E) Games

The NFT gaming ecosystem has expanded significantly, with platforms like Decentraland and Cryptovoxels introducing players to the possibilities of the Metaverse. Decentraland, for example, is a virtual world built on the Ethereum blockchain where users can buy, sell, and trade virtual land, properties, and in-game assets. Similarly, Cryptovoxels and Sandbox NFTs—introduced in 2020—gave birth to new virtual experiences where players could purchase land, digital assets, and much more.

But the real game-changer is how Play-to-Earn (P2E) mechanics are transforming the gaming industry. Thanks to the development of virtual reality (VR) and devices like the Oculus headset, players are now able to immerse themselves in virtual worlds akin to Ready Player One, where they can buy in-game items, land, characters, and skins—all as NFTs.

These in-game items, including virtual properties, rare assets, and special skins, can be purchased on platforms like OpenSea and LooksRare. Upon purchasing an NFT, ownership is transferred to the buyer's Ethereum address, and they can use it within the game. The beauty of NFTs in this space is the ability to resell items to others. If you grow tired of your in-game asset or find something better, you can list your NFT for sale on secondary marketplaces, allowing you to recoup some of your investment or even make a profit.

The Metaverse: A New Frontier for Gaming and Digital Ownership

To paint a clearer picture, imagine playing a game like Grand Theft Auto (GTA) Online. In GTA, players can purchase virtual assets like properties, cars, planes, warehouses, and boats. However, there's a catch: while you may spend significant amounts of money on these items, you can't truly monetize them. These assets exist within the game, and you can't sell or trade them outside the platform. The entire experience remains confined to the digital world of the game, with no ability to convert your assets into real-world value.

Now, envision a future where games like GTA, Call of Duty, or Fortnite evolve into Play-to-Earn games. Players can still buy in-game items—properties, vehicles, weapons, skins, and more—but now they hold tangible ownership. This means that the items you purchase with real money are not just confined to your game account. They exist as NFTs that you own and control, and if you decide you no longer want a specific item, you can sell it on an NFT marketplace to another player. This introduces a true economy within the gaming world.

In traditional games, like Fortnite, many players spend thousands of dollars buying skins and rare items during each season. These digital goods are often non-transferable and only hold value within the game. Once you purchase a limited-edition skin, you own it, but it's useless beyond the game—until now.

With the NFT revolution in gaming, you can now take ownership of your in-game items. If you're tired of a certain skin or item, you can sell it to another player or use it in a different game that supports cross-game assets. This gives players a unique way to monetize their investments in digital items, making gaming more financially rewarding.

A New Era for Gamers and Investors

The integration of NFTs into the gaming industry introduces a revolutionary concept: ownership and tradeability of in-game assets. In addition to enhancing the gaming experience, these new features open up opportunities for both gamers and investors. Whether you're purchasing virtual land in the Metaverse, collecting in-game skins, or trading rare items, NFTs provide a new layer of value that was previously unattainable in traditional gaming systems.

In the world of P2E games, players now have the potential to earn cryptocurrency by participating in the game, completing tasks, and trading in-game NFTs. This model incentivizes active participation, with players gaining tangible rewards for their time and effort, unlike traditional gaming where players spend money without any chance of recouping it.

Furthermore, as more games adopt NFTs and integrate virtual economies into their platforms, the value of these digital assets continues to rise, further blurring the lines between gaming and investing. With the ability to buy, sell, and trade items with ease, NFTs offer a wealth of opportunities for players, collectors, and investors alike.

The Future of NFTs in Gaming and the Metaverse

As the NFT market continues to evolve, we're likely to see more immersive, Metaverse-style games with integrated NFT economies. From land sales to in-game assets, NFTs will provide a way for players to truly own their digital assets and monetize their efforts. These advancements open up exciting possibilities for both creators and players, allowing for an entirely new way to experience, invest in, and profit from the digital world.

The Play-to-Earn model, combined with NFTs, offers a new frontier for the gaming industry. It's no longer just about playing for fun—it's about playing to earn, invest, and build wealth. As the technology and user experience improve, the future of NFTs in gaming promises to be a game-changer, both for the industry and for the players who take part in it.

The NFT revolution is just getting started. Now is the perfect time to get involved, whether you're an avid gamer, a digital creator, or an investor looking to explore this exciting new realm. The possibilities are limitless, and the opportunities are waiting to be seized.

How to Make Money with NFTs and the Power of Royalties

As the NFT space continues to grow, a key feature that makes NFTs even more enticing is the ability for creators to earn royalties on their digital assets. Once you mint and sell an NFT, you retain the potential to earn royalties each time your creation is resold in secondary marketplaces. These royalties provide a consistent, passive income stream for creators, enabling them to benefit from their work long after the initial sale.

How NFT Royalties Work

NFT royalties are a game-changing feature for creators. These royalties are embedded within the smart contract of the NFT, allowing the original creator to receive a percentage of every resale. Typically, royalties range from 5-10%, but creators have the freedom to set their own royalty percentage during the minting process. Once an NFT is listed for sale in marketplaces like Opensea or LooksRare, the royalties are automatically transferred to the creator each time the asset is sold in the secondary market. This means the creator can keep earning money on their creation each time it changes hands, for as long as it continues to resell.

In essence, NFT royalties provide a form of passive income that is built into the very structure of the NFT. They can grow overtime, continuously adding value for the creator without the need for any further work. This makes NFTs a powerful tool not only for artists but also for anyone who wants to sell digital goods and earn long-term rewards.

Passive Income Through Royalties

One of the most exciting aspects of NFT royalties is their perpetual nature. As long as your NFT is being resold, you continue to benefit from it. Imagine creating a digital artwork, minting it as an NFT, and then earning royalties from each resale that happens thereafter. This model is incredibly attractive, especially for content creators, musicians, visual artists, and even digital entrepreneurs. By embedding royalties in the smart contract during the minting process, creators make sure they don't miss out on future earnings as their work becomes more valuable over time.

For example, if you create a piece of artwork and set a 10% royalty, each time it's sold in the secondary market, you'll automatically receive 10% of the sale price. This mechanism allows creators to continually profit from their work, regardless of how many times it's sold in the future.

It's also worth noting that these royalties can be tracked on blockchain explorers like Etherscan.io, giving full transparency to both the buyer and the seller.

Who Can Make Money with NFTs?

The beauty of NFTs is that they are accessible to virtually anyone with a creative idea or a digital asset to sell. It's not just artists and musicians that can benefit from NFTs. In fact, the potential for earning money through NFTs extends to a wide range of industries. Here are some examples of who can profit from NFTs:

- **Content Creators**: Writers, designers, and creators of digital media can sell their work as NFTs, earning royalties with each subsequent sale.
- **Musicians**: Artists can tokenize their music as NFTs, offering fans a chance to own exclusive tracks or albums and receiving royalties every time the music is resold.
- **Visual Artists**: Digital and traditional artists can sell their artwork as NFTs, keeping royalties as their work circulates through secondary markets.

- **Real Estate**: Virtual real estate in Decentraland, Cryptovoxels, and other Metaverse worlds is being sold as NFTs, giving owners the potential to resell for a profit.
- **Retailers**: Brands like Amazon, Walmart, and other companies can issue limited-edition digital products as NFTs, adding value for customers and earning royalties from resales.

- **Car Dealers, Restaurants, and Other Businesses**: NFTs offer an innovative way for businesses to tokenize products and services, offering customers the ability to own a part of the brand digitally while providing royalties for the business each time the NFT changes hands.

Important Considerations

While it's tempting to dive into the NFT space with the hopes of making significant profits, it's essential to remember that success is not guaranteed. The NFT market is highly saturated, and the vast majority of NFTs do not sell for large sums. Therefore, creators must be strategic and thoughtful about what they create, and buyers should do their research before investing.

If you're planning to create and sell NFTs, it's crucial to ensure that your content has value, whether it's an art piece, collectible, or digital asset. As with any investment, it's important to understand the market dynamics and the rarity of the NFT to make informed decisions.

How to Start Researching and Buying NFTs

If you're interested in buying or listing NFTs, there are several useful resources that will help guide you on your journey. To get started, it's important to understand both the marketplaces where NFTs are sold and the rarity of the assets you're considering purchasing.

Here are some trusted websites you can use to begin your NFT journey:

- Opensea: One of the largest NFT marketplaces for buying and selling a variety of digital assets.
- LooksRare: Another NFT marketplace with a focus on community-driven features and rewards for users.
- Rarible: A decentralized NFT marketplace for artists, creators, and collectors.

- SuperRare: A platform focused on showcasing high-quality artwork as NFTs.
  Foundation: A platform for artists to showcase and auction their NFTs.

  Mintable: A marketplace for creators to mint and sell NFTs easily.

To understand the rarity and value of NFTs, you can use resources like:

- CryptoSlam: A platform that tracks NFT sales and helps you assess the value of various projects.

- NonFungible: Offers data and insights into the NFT market and helps you track the performance of various NFTs.

- NFTGo: Provides in-depth analytics and market data for NFTs.

- Icy.Tools: A popular website for tracking NFT trends, sales, and analytics.

By utilizing these platforms, you can research NFT trends, track your investments, and make informed decisions about which NFTs to buy or sell.

NFTs represent a new era of digital ownership and monetization. By creating NFTs, you can earn royalties for as long as your creations are traded on secondary markets, making it a powerful tool for generating passive income. Additionally, anyone—whether they're an artist, entrepreneur, or collector—can participate in the NFT economy and potentially make money by buying and selling NFTs.

However, it's essential to approach the market with caution, do your research, and understand the risks involved. With the right strategy, NFTs can provide a lucrative avenue for creators and investors alike. Now is the time to start exploring this exciting new world of digital assets and unlock the potential for passive income through NFT royalties.

- **https://opensea.io/**

- **https://nftgo.io/**

- https://marketplace.axieinfinity.com/
- https://www.larvalabs.com/cryptopunks
- https://nbatopshot.com/marketplace
- https://rarible.com/
- https://superrare.com/
- https://foundation.app/
- https://niftygateway.com/
- https://mintable.app/
- **https://www.thetadrop.com/**
- **https://magiceden.io/**
- looksrare.org/
- https://solanafloor.com/
- https://solanart.io/
- https://digitaleyes.market/
- https://www.metaplex.com/
- https://alpha.art/
- **https://Exchange.Art**
- https://www.supadrop.com/

- **https://zora.co/**
- https://nftify.network/
- **https://crypto.com/us**
- **https://www.binance.com/en/nft/home**

How to Get Started with NFTs and Make Profits

If you're looking to start your journey in the NFT space, there are several reliable and credible platforms where you can begin purchasing your first NFTs. The most popular marketplaces are OpenSea and LooksRare, where a large portion of the trading volume takes place. But remember, always Do Your Own Research (DYOR) to ensure you're making well-informed decisions.

Along with purchasing NFTs on trusted platforms, NFT evaluation and rarity tools are essential to help you stay ahead in the game. These platforms are the hidden gems that help you find those rare NFTs, track big-money wallets, and make better investment choices. With the right tools and analytics, you can stay on top of trends, follow whale wallets (big investors), and evaluate the rarity of assets—critical aspects in making informed investment decisions.

The Importance of Information in NFT Investing

Just like in any other crypto space, having access to accurate information is crucial for your success as an NFT investor. Whether you're a beginner in the NFT world or an active collector, these analytics tools help you assess the rarity and availability of assets. After all, you probably want to find the next Bored Ape, Mutant Ape, or CryptoPunk, right?

Here's a list of websites and tools you can start using to evaluate the rarity and performance of NFTs:

- Icy Tools
- NFT Nerds
- Dapp Radar
- Trait Sniper
- Opensea Activity (Put on Ethereum Chain)
- Rarity Tools
- Rarity Sniper
- Rarity Sniffer
- Bitdegree NFT Tracker
- Moby.gg
- NFT Drops Calendar
- Upcoming NFT
- Coinrivet NFT Calendar
- Crypto.com NFT Collections
- Nansen
- CryptoSlam
- NFT Bank
- NFT Stats

These websites are your arsenal for analyzing trends, tracking whale activity, and understanding the rarity of different NFTs. With the right data, you can make smarter choices when purchasing NFTs, ensuring you're investing in projects with real potential.

How to Make Money with NFTs

The big question: How can you make money with NFTs? Let me share a simple game plan for getting started and profiting from NFTs. Let's take the example of Tracy, who has $1,000 to spend on. NFTs Here's how she could break it down:

- Tracy buys three NFTs for $300 each, totaling $900 spent (plus gas fees).
- She sells the first NFT for 3X what she paid for it, making $900 profit on that one NFT.
- The second NFT is listed for 5 ETH on Opensea (equivalent to $12,756 at the time of writing), and Tracy successfully sells it.
- Tracy holds onto her third NFT, which is a rare one in the top 10 of a collection of 10,000 NFTs. She's waiting for the floor price to hit an all-time high, planning to sell it for over 10 ETH (approximately $25,000).

In this case, Tracy's $1,000 investment turned into over $13,000 in profit—without including the potential of her third NFT going up in value. This is a real example of how strategic NFT investments can yield significant returns if you research the right projects, and the right NFT marketplaces and tools.

The Power of NFT Communities and Use Cases

To maximize your chances of success in the NFT space, it's essential to recognize that community-making it much more likely to increase in value over time. Simply flipping JPEGs may bring short-term profits, but NFT projects with strong communities and real-world applications tend to stand the test of time.

The Future of NFTs: Endless Possibilities

The NFT space is evolving rapidly, and there are endless possibilities for what can be turned into an NFT—from a simple pen drawing to a valuable painting, everything can become a non-fungible token. With the right knowledge, tools, and strategies, you can capitalize on this emerging market.

Start Your NFT Journey Today!

Now that you have the knowledge and tools, what's stopping you from diving into the world of NFTs? If a 14-year-old can make millions from selling NFTs, there's no reason why you can't too! The NFT market is still young and brimming with opportunity, and the key to success lies in doing your due diligence, evaluating rare assets, and joining communities with real potential.

Start listing your NFTs, use the tools provided, and start making money today. Whether you're an artist, a collector, or an investor, the NFT revolution is waiting for you. So, what are you waiting for? The future of digital ownership is here!

Chapter 13

Metaverse: Are We Moving Towards Virtual Lifestyle?

Do you remember the movie *Ready Player One*, or perhaps the iconic *Matrix* trilogy? Maybe *Tron*, *Replicas*, *Blade Runner 2049*, *Ghost In The Shell*, *Creative Control*, *The Congress*, *The Zero Theorem*, *Total Recall*, *Source Code*, *Inception*, *The Island*, *Minority Report*, *The Cell*, *The Thirteenth Floor*, *eXistenZ*, *Lawnmower Man 2*, *Strange Days*, *Hackers*, or even the movie *Virtuosity* starring Denzel Washington? Or maybe *Johnny Mnemonic*, *Disclosure*, *Brainscan*, the 1993 movie *Demolition Man* starring Wesley Snipes, Sandra Bullock, and Sylvester Stallone, or *Arcade*, *Brainstorm*, and *Videodrome*, which first aired in 1983? All these movies shared a common theme: they depicted a world in where people escaped from the mundane to Virtual Reality (VR).

Now imagine a world, maybe the world you actually live in now, where you work a 9-5 job, and you do the same routine every day—what if you could escape for a few hours and immerse yourself in a world where you can communicate, make money, conduct business meetings, and experience life differently, without the need to drive for hours through a New York traffic just to reach the office?

As kids, we often thought that the technological concepts we saw in these movies would never become a reality, but the truth is that we've been paving the way these ideas for years. I remember when I was a child, my grandparents gave me an amazing Christmas present: a *Virtual Reality Ninja Game*. It was a headset, and through it, you'd enter a virtual world where ninjas would attack, and you'd defend yourself using special gloves and foot straps. This game, which I played in the early 2000s, marked my first real interaction with virtual gaming.

Looking back, I can see how this immersive experience sparked my love for virtual worlds. Being able to step into a different world, where I could become someone else, provided an escape, a fantasy that, as a kid, drew me to video games. Even today, I still play my fair share of games. The impact of VR on my life is clear, and I know it will similarly affect my children as they grow up. But what about those who have delved deeper into virtual reality? What are the positive and negative impacts of this technology?

In this book, we aim to educate not only on the history of where I see the cryptocurrency markets heading, but also on the potential of virtual games and the future of the Metaverse. It's essential to approach these technologies with an ethical mindset, ensuring that they don't cause harm to the environment, humanity, or animals. They should help society rather than tear it down.

The Rise of Virtual Reality

One of the biggest trends in the cryptocurrency space—aside from decentralized finance—is the rise of virtual reality, and more specifically, the Metaverse. I want to clarify that this journey toward the Metaverse has been ongoing for years, not just because Facebook rebranded as *Meta* on October 28, 2021.

The concept was built long before that, and we can trace its origins to when the internet began in 1991.

The year1991 marked the advent of the internet, which could easily be considered the starting point of the Metaverse. But there are other milestones we should point out that have led to this groundbreaking development. History repeats itself, and I believe that the future is a continuous thread, with no clear beginning or end. The breakthroughs we see today in technology are built upon the framework of past discoveries. Looking back, we can now clearly see how the historical steps taken before were pivotal in forming what we now know as the Metaverse.

The word "Metaverse" itself is a portmanteau of the Greek word "meta," which means "beyond," and the English word "universe." In Greek, "meta" is often used as a prefix to mean "after" or "beyond," and this concept carries over into the English language. For example, the term "metadata" refers to information that describes other data, and "metaphysics" refers to what lies beyond the physical. When we apply this idea to the Metaverse, we're talking about a virtual universe that exists beyond the physical one, where digital interactions and experiences take place in real-time.

The Metaverse, as we envision it today, is a massively scaled platform built on blockchain technology. It's an interoperable network of real-time rendered 3D virtual worlds, where in-game avatars and items exist and can be experienced simultaneously by a virtually limitless number of users. This experience creates a sense of presence, allowing individuals to interact in virtual spaces, while maintaining continuity in data such as identity, history, entitlements, and even financial transactions.

The Metaverse is no longer a distant fantasy—it is becoming a reality, and it's important to understand both its potential and the risks that come with it. As we move forward, we must consider how we can use this technology ethically and responsibly. But one thing is clear: the future of the Metaverse, like the internet before it, will continue to shape how we live, work, and connect with each other.

To better understand the Metaverse, let's break down its key properties:

- Massively scaled
- Interoperable
- Real-time rendered
- 3D virtual worlds
- Synchronous

- Persistent
- Unlimited number of users
- Individual sense of presence
- Continuity of data
- In-game avatars
- In-game items (such as land and skins)

A Brief History of the Metaverse

On August 6th, 1991, Tim Berners-Lee made the first collaboration on the WorldWideWeb, marking the official birth of the internet. The following year, in 1992, science fiction writer Neal Stephenson coined the term "Metaverse" in his novel *Snow Crash*. The concepts introduced in this book have influenced the Metaverse we are developing today—where humans, as avatars, interact with each other and software agents in a three-dimensional virtual space that mimics the real world.

Some of the key ideas Stephenson envisioned for the Metaverse, which are reflected in today's digital landscape, include:

- The Metaverse is three-dimensional.
- It serves as a metaphor for the real world.
- Users can access the Metaverse using goggles, much like today's VR headsets.
- Users experience the Metaverse from a first-person perspective.
- Virtual avatars in the Metaverse are customizable to some degree.

Early Virtual Reality Innovations

In 1992, Jason Lanier, CEO of the virtual reality headset, VPL Research, introduced Virtual Reality Musical Instruments (VRMI) for the first time during his performance "Sound of One Hand" in Chicago Jason used a VR headset to present musical instruments designed within virtual reality, including the EyePhone headset, DataGlove, and a tactile hand interface. These instruments, like the CyberSax, Rhythm Gimbal, and CyberXylo, marked early attempts at blending the virtual and physical worlds in a creative, interactive way.

In 1993, the term and concept of "Proof of Work" (PoW) emerged in the context of computer security as a way to prevent email spamming. Over time, PoW evolved and became a primary method of verification, particularly in digital currency mining. This method relies on computing power to verify transactions on a blockchain network.

In 1998, Wei Dai introduced the concept of a decentralized and distributed cryptocurrency called "BMoney." Although B-Money was never launched, its ideas were influential in the creation of Bitcoin, which emerged in 2009. One key feature that emerged from this concept was the use of Proof of Stake (PoS), an alternative mining algorithm that relies on the holding of a particular cryptocurrency rather than relying on massive computational power.

In 2000, research firm Gartner coined the term "Supranet," referring to the digital counterpart of both the virtual and physical worlds in the Internet age. The article, titled *"Beyond the Internet: The 'Supranet,'"* outlined this idea of an interconnected, digital ecosystem.

In 2001, the massively multiplayer online role-playing game (MMORPG) RuneScape was released, offering a glimpse into the potential of virtual worlds. Fast forward to 2002, when the concept of the "digital twin" was introduced. This idea, first proposed by Michael Grieves and presented at a conference by the University of Michigan, referred to the digital representation of physical objects, setting the stage for future applications in industries like manufacturing and product lifecycle management.

The year 2003 marked the creation of *Second Life*, an online virtual world developed by Philip Rosedale and his team at Linden Lab. Though its bandwidth limitations and high latency hindered the experience, Second Life remains one of the early pioneers in virtual environments, with millions of users spending several hours a day in this immersive digital world.

In 2006, *Roblox* was introduced as an online platform where users could create and play games developed by others. This platform gained considerable traction, particularly during the pandemic in 2020, and became one of the highest-grossing games of that year.

On January 3, 2009, amidst the aftermath of the 2008 financial recession, Satoshi Nakamoto unveiled Bitcoin, a revolutionary concept in digital currency, underpinned by blockchain technology. While some had proposed blockchain ideas earlier, Nakamoto's implementation introduced a usable form of the technology that would later revolutionize how digital transactions were handled. On the same day, the mysterious Nakamoto mined Bitcoin's "genesis block" (block number 0), earning a reward of 50 bitcoins.

This moment marked the birth of Bitcoin and, by extension, the blockchain era.

In 2010, the world was introduced to "Play-to-Earn" technology, and one of the early manifestations of this concept was the Gacha game. Popular in Japan, the Gacha games were inspired by capsule-toy vending machines, where players could spend real-world money to earn virtual items in a randomized fashion, adding a new layer to the concept of virtual economies and gaming.

Through the evolution of these various concepts, the seeds of the Metaverse and virtual reality as we know them today were planted. From early virtual worlds and decentralized financial technologies to the advent of gaming platforms that blurred the line between entertainment and real-world value, the landscape was gradually evolving into what we now recognize as a potential future of digital interaction— interconnected, immersive, and decentralized.

One of the earliest examples of blending real-world value with gaming was seen in the concept of "toy vending machines" in games like *MapleStory*. Players would earn currency that they could use to participate in random draws for in-game items based on rarity. The goal was often to collect an entire set of items to earn powerful in-game rewards. This early form of randomized, collectible assets laid the groundwork for the later emergence of NFTs, or Non-Fungible Tokens, in the space.

An NFT is a unique digital asset that represents ownership of a specific item, unlike fungible tokens like Bitcoin, which are interchangeable. The concept of NFTs dates back to December 2012 with the creation of "Colored Coins," which added metadata to Bitcoin, making it unique. This concept was spearheaded by a young Vitalik Buterin, who would later go on to co-create Ethereum. By 2012, *RuneScape*, a massively multiplayer online role-playing game (MMORPG), had reached an estimated 200 million users, showcasing the increasing scale and importance of virtual worlds.

In 2013, the virtual reality space took a major leap when Oculus debuted its first prototype VR headset to Kickstarter backers. This laid the foundation for immersive virtual experiences, which would later become a crucial part of the Metaverse vision. The following year, in 2014, Zwift, an immersive cycling game, was released, pushing the boundaries of virtual reality integration in sports and exercise.

In parallel with these technological advancements, Vitalik Buterin's journey into the world of blockchain and decentralized applications began with support from Peter Thiel. Thiel's Thiel Fellowship, which offered $100,000 grants for students under 22 to leave school and pursue innovative ventures, played a key role in allowing Buterin to develop Ethereum. Launched in July 2015, Ethereum introduced the concept of *smart contracts*. Originally proposed in the 1990s by Nick Szabo, smart contracts are digital agreements executed on a blockchain. Since Ethereum's implementation, smart contracts have become integral to the development of decentralized applications (DApps) and play-to-earn games, where users can engage in economic activities within virtual worlds.

By 2015, Decentraland was introduced as a virtual reality platform that allowed users to own and trade "land" in a virtual world, using a proof-of-work algorithm. This was one of the first instances of a virtual space where real-world value could be created through digital assets like land. Decentraland's success highlighted the potential of NFTs in virtual real estate, with some virtual properties being sold for over $100,000, marking the beginning of a new era in digital ownership.

In 2016, a significant shift occurred as virtual worlds began merging with the real world in new and exciting ways. These developments, along with the rise of decentralized technologies and the growing popularity of virtual worlds, set the stage for what would become the Metaverse—a truly interconnected digital universe where users could live, work, and play. The Metaverse, powered by blockchain, NFTs, and immersive VR experiences, promised to redefine the way we interact with digital spaces and each other. This vision of a decentralized, virtual future continues to unfold today, as new technologies and platforms emerge, further blurring the lines between the physical and virtual realms.

The concept of virtual reality and the Metaverse continued to evolve with the introduction of Pokémon Go in 2016. This groundbreaking mobile game utilized GPS technology to allow players to capture and battle virtual creatures in real-world locations, blending the virtual world with the physical one in real-time. It marked a major milestone in augmented reality, attracting millions of users globally, and offered a glimpse into how the Metaverse could bridge the digital and physical realms.

Later in 2016, The DAO was launched through an Initial Coin Offering (ICO), setting the record for the largest crowdfunding campaign of its time. The DAO, built on Ethereum's blockchain, aimed to be a decentralized venture capital fund. However, in June 2016, a vulnerability in the DAO's code was exploited, resulting in the theft of one-third of its funds. The collapse of The DAO served as a critical learning experience, highlighting the potential risks and challenges of decentralized systems. Despite its failure, the concept of decentralized autonomous organizations (DAOs) continues to thrive. DAOs now govern various blockchain projects, including NFTs, and serve as a foundational element in the creation of future Metaverse companies, where participants collectively make decisions and record transactions on the blockchain.

In 2017, the gaming industry saw the rise of battle royale games, inspired by Koushun Takami's 1999 novel *Battle Royale* and later adapted into the popular movie *Battle Royale*, which served as the inspiration for the *Hunger Games*. The *battle royale* format introduced 100 players battling simultaneously in an online arena, with the last player standing earning the title of the "Winner, Winner, Chicken Dinner." PlayerUnknown's Battlegrounds (PUBG) launched in March 2017 as one of the first battle royale games, followed by the hugely successful *Fortnite*. Fortnite's success not only popularized the battle royale genre but also introduced many gamers to the concept of virtual currency. In-game purchases, including skins and items, became a driving force for the game's success, with Fortnite earning over $9 billion in revenue and amassing a user base of 350 million players.

Parallel to these developments, the concept of the Metaverse continued to take root in pop culture. In 2011, Ernest Cline's novel *Ready Player One* envisioned a future where people would escape to a vast virtual universe to experience everything from gaming to social interaction, and even work. The book, released as a movie in 2018, introduced a new generation to the possibilities of virtual reality. *Ready Player One* helped bring the Metaverse concept to the forefront, sparking imaginations and fueling the drive for real-world virtual worlds to come to life.

Later in 2018, another significant milestone occurred in the cryptocurrency space with the creation of the Dai Stablecoin. Unlike cryptocurrencies that lack stability or are backed solely by other cryptocurrencies, Dai was pegged to a combination of fiat currency and other crypto assets, such as USD Coin (62%) and Ethereum (ETH) (29%). This added a new element to the volatile crypto universe by providing a more reliable and less volatile currency, making it especially suitable for decentralized finance (DeFi) applications.

In the same year, the rise of Decentralized Exchanges (DEXs) opened up new opportunities for peerto-peer transactions without the need for Know Your Customer (KYC) processes mandated by centralized exchanges. This shift empowered individuals to trade cryptocurrency assets directly, based on smart contracts, without relying on a central authority. DEXs remain a pivotal part of the crypto ecosystem, providing an alternative to traditional finance systems and fostering decentralized financial activities.

Axie Infinity, a popular NFT-based virtual reality game, was also introduced in 2018, setting the stage for the convergence of gaming, NFTs, and the Metaverse. The game, set in a mythical world of animal husbandry, became a massive success. By mid-2021, it had achieved the highest combined volume of NFT purchases across play-to-earn game platforms. This shift in the gaming landscape mirrored the rise of the Metaverse, with players earning real-world value through in-game assets and tokens.

By 2019, the Metaverse had further matured. The creators of Sandbox and Dual Universe innovated by enabling 30,000 users to interact within a continuously live and synchronous single shard, simulating an early vision of the Metaverse. That same year, Roblox, a platform that allowed users to create and play games developed by others, saw its economy expand dramatically. In-game purchases on Roblox reached $110 million, with over half of its users customizing avatars each month, keeping money circulating within its virtual universe economy.

The year 2020 was a year of dramatic shifts, spurred by the global pandemic. With the onset of lockdowns and quarantines, people were forced to adapt to new ways of socializing and connecting. The term "social distance" became ingrained in society, and traditional ways of interaction gave way to digital spaces. The Metaverse became the go-to platform for millions—particularly young people, gamers, and those seeking economic opportunities in the digital space.

In this rapidly changing environment, the Metaverse began to take on the role of a rebellion against traditional institutions, providing a refuge for those seeking freedom from the physical world's limitations.

The pandemic accelerated the adoption of Decentralized Applications (DApps), which powered decentralized finance (DeFi) and other blockchain-based services. DApps enabled transparent, open-source platforms for gaming, exchanges, and financial transactions. While DApps are often hailed as an emerging trend, the first DApp—Bitcoin—was actually created more than a decade ago, and today, many applications on the blockchain are built on similar principles.

In 2020, the world also saw a surge in virtual concerts as artists adapted to the new digital reality. In April, *Fortnite* hosted a groundbreaking virtual performance by Travis Scott and Marshmello, attracting nearly 30 million attendees. This marked a pivotal moment in the Metaverse's cultural impact, where digital platforms became viable venues for entertainment and performance.

That same year, the Solana blockchain introduced a decentralized application that provided an alternative to Ethereum's proof-of-work (PoW) algorithm. Solana's proof-of-stake (PoS) consensus mechanism, along with its innovative "proof of history" feature that inserted timestamps into its transactions, clarified and simplified block ownership, making it a faster and more efficient platform for decentralized applications.

The year 2020 was a perfect storm for the Metaverse's growth—the convergence of the pandemic's social distancing measures, decentralized finance, the rise of virtual worlds, and blockchain technologies. With the world increasingly moving toward digital spaces for interaction, entertainment, and financial transactions, the Metaverse's potential to reshape our lives became clearer than ever.

The developments in virtual reality, cryptocurrency, decentralized exchanges, and blockchain technology set the stage for the next era of digital interaction—one where the boundaries between the physical and virtual worlds would continue to blur.

In 2020, *Nintendo* launched *Animal Crossing: New Horizons*, an idyllic virtual world where players interact with flora, fauna, and villagers. It was one of many significant events that helped shape the Metaverse. That year, *Alien Worlds*, a wildly popular decentralized application (dApp), was created. It featured a multi-Metaverse, interplanetary scenario, where NFT characters interacted in a decentralized autonomous organization (DAO) to mine tokens and perform various tasks. By 2021, *Alien Worlds* had reached over 2.5 million users. Its significance extends beyond its popularity; the game serves as an educational tool, teaching players about the principles of cryptocurrency and the benefits of crypto mining.

In 2021, *Epic Games* secured $1.78 billion in funding to advance its vision of building Metaverse experiences. As the Metaverse gained momentum, global governments and social media giants like Facebook began exploring ways to incorporate blockchain technology into their own Metaverse platforms.

The concept of the Metaverse, as popularized by *Ready Player One*, is now a reality. The ability to earn real-world currency from playing games, owning virtual land that generates passive income, and engaging in virtual worlds has become achievable.

Now that we've covered the history of how virtual reality platforms and the Metaverse have evolved, let's explore the opportunities and benefits they present:

Benefits of the Metaverse

- **Overcoming obstacles**: The Metaverse can help individuals with disabilities perform activities that are difficult or impossible in the physical world.
- **Enhancing creativity**: It fosters creativity and imagination, allowing individuals to explore new ideas and build in ways never before possible.
- **Traveling without moving**: People can experience different places around the world virtually, without physically traveling.
- **Earning income**: The Metaverse provides opportunities to make money online, freeing people from traditional 9-to-5 jobs and promoting a "work smarter, not harder" mentality.
- **Increased technological literacy**: It provides a platform to develop essential skills in a techdriven world.
- **Connecting with others**: The Metaverse enables social connections without the awkwardness or barriers of in-person interactions.
- **New job opportunities**: The Metaverse is opening doors for careers and professions that didn't exist before.
- **Connecting with loved ones**: Virtual spaces allow us to meet and spend time with family and friends, no matter the distance.
- **More educational possibilities**: It opens doors for innovative learning experiences, both formal and informal.
- **Self-expression**: It gives individuals a unique platform for expressing themselves creatively.

- **Remote work solutions**: The Metaverse can enhance remote work experiences by providing virtual spaces for collaboration and productivity.
- **Healthcare advancements**: Virtual platforms are becoming valuable tools for healthcare professionals in training and patient care.
- **Exciting online games**: Gamers now have an immersive environment for more dynamic and engaging experiences.
- **Virtual real estate tours**: Industries like real estate benefit from virtual tours, offering a whole new way to explore properties.

While these advantages are compelling, there are also ethical concerns to consider. The Metaverse could lead to people becoming detached from reality, using it as an escape from real-life challenges. It's important to approach this technology with caution and ensure that, as we dive deeper into virtual worlds, we maintain a strong connection to nature and the physical world.

The Metaverse should be viewed as a tool for enhancement, not an all-consuming alternative to real life. Don't lose sight of the world around you; virtual experiences can be enriching, but the beauty of life should not be overshadowed by the allure of the digital realm.

The Evolution of the Metaverse: A New Digital Economy

The Metaverse promises to redefine our digital experiences, constantly evolving to meet the needs of

its users. Augmented Reality (AR) and Virtual Reality (VR) have been at the forefront of this transformation, with both technologies gaining significant attention from major tech companies. These companies have been investing heavily in VR and AR equipment to support the development of futuristic Metaverse projects, as we transition into the era of Web 3.0.

One crucial component driving the Metaverse is the Internet of Things (IoT). IoT acts as the bridge between our physical world and the digital one, enabling seamless data exchange between sensors and devices. In the context of the Metaverse and play-to-earn games, IoT collects data from the physical world and transfers it into virtual environments. This enhances the accuracy of digital representations, enabling real-world objects to function accurately within virtual spaces.

For instance, IoT data feeds can monitor and adjust the behavior of objects within the Metaverse based on real-time environmental changes. By combining IoT with Artificial Intelligence (AI) and Machine Learning (ML), developers can optimize data management, creating more responsive and immersive virtual experiences. This synergy of technologies is taking the Metaverse to new heights.

With these technological advancements in mind, many are eager to understand how they can participate in the burgeoning Metaverse economy, specifically through play-to-earn games. These games not only offer entertainment but also present an opportunity to earn real cryptocurrency. Some players are making as much as $16 per hour just by participating in these games.

Here are some of the most popular play-to-earn games you can start playing today to earn cryptocurrency:

1. Axie Infinity – An NFT game for Pokémon fans.
2. The Sandbox – The best NFT game for Minecraft players.
3. Gods Unchained – An NFT game for fans of Magic: The Gathering.
4. DeFi Kingdoms – An NFT game for pixel art fans.
5. Gala Games, Town Star – A strategy-based NFT game.
6. Splinterlands – The Hearthstone of NFT games.

7. Sorare – Fantasy football; great for fans of Panini stickers.

8. Treeverse – An NFT game for mainstream gamers.

9. Bridgeworld (TreasureDAO) – For fans of old-school text adventure games.

10. Star Atlas – An immersive NFT game for space exploration fans.

11. Parallel – A social NFT game for gamers who enjoy interaction.

12. Wizard and Dragons – A fun 8-bit RPG for fantasy lovers.

13. Thetan Arena – A multiplayer online battle arena game.
14. Forest Knight – An RPG involving knight battles against evil forces.

15. Farmers World – A farming game with tools to grow and harvest crops.

16. CryptoBlades – A role-playing game in the crypto space.

17. Plant vs Undead – A tower defense game where players purchase SEEDS to play.

These are just a few examples of the many games available where you can earn cryptocurrency while having fun. The idea of spending a few hours playing a game, completing missions, or selling in-game items to earn real digital assets is transforming how we perceive work and leisure. Cryptocurrency earned in these games can easily be transferred from your wallet to other digital platforms, opening up a world of financial opportunities.

A New Generation of Wealth

I firmly believe that those who get involved in the Metaverse now—whether by creating content on the blockchain or diving into play-to-earn games—are laying the foundation for what will be a generational shift in wealth. Already, young people, some as young as 12-14, are making millions of dollars by creating and selling NFTs, building financial freedom and wealth from digital assets.

This new digital economy has the potential to create a new generation of millionaires and billionaires like we've never seen before. This is why I titled this book *The Biggest Transfer of Wealth in the Age of Cryptocurrency*. The next wave of wealth creation is happening right now, and it's poised to benefit Millennials and Generation Z in ways we have yet to fully comprehend.

As we witness the rise of blockchain technology, NFTs, and the Metaverse, we're seeing the birth of a new digital frontier—one that promises to democratize wealth creation, empower individuals, and reshape our economic landscape for generations to come.

The Metaverse and Play-to-Earn: The Biggest Transfer of Wealth in the Digital Age

The impact of the Metaverse, NFTs, and play-to-earn games has been significant, especially among younger generations. According to a Harris Poll, 7 out of 10 Generation Z and millennials express interest in engaging with the Metaverse. Another survey revealed that 38% of Generation Z believes the Metaverse will become the next big thing, while 48% of millennials, aged 25 to 40, share the same sentiment. These numbers indicate just how far-reaching and rapidly growing the concept of the Metaverse has become.

The buzz around the Metaverse has only intensified following Facebook's rebranding as Meta, signaling a clear shift toward virtual worlds. Corporate giants such as Nike, Adidas, Ferrari, Aston Martin, McDonald's, and Subway are already preparing to dive into the virtual space with their own projects. Wall Street analysts have become increasingly optimistic about the future of the Metaverse, with many predicting that young people—already accustomed to gaming and spending time online—will naturally embrace these virtual realities.

This growing trend points to an exciting future where the Metaverse will cater to a wide range of industries, each introducing its own unique version of virtual spaces and experiences.

The question now is: What's stopping you from getting started? Imagine earning more than $16 a day by playing your favorite games, all while having fun in a virtual world. With a VR headset, the Metaverse is just a few steps away. So why wait? Dive into this incredible digital frontier today—I'm looking forward to seeing you in the Metaverse!

Chapter 14

Web 3.0: Unplugging From the Matrix

The internet has come a long way since the days of Web 1.0, and the journey has brought us to the dawn of a new era—Web 3.0. Before we dive into that, let's look at the evolution of the internet and how it's transformed from a simple information-sharing tool into the complex, interactive, and decentralized space we are now exploring.

Web 1.0: The Early Days of the Internet

Web 1.0, the first generation of the World Wide Web, is often referred to as the "read-only" web. During this period, the internet was essentially a static space where users could access information, but they couldn't interact with or contribute to it in any meaningful way. Websites were built using simple HTML and displayed static pages that provided content but didn't allow for dynamic interaction or updates.

Characteristics of Web 1.0:

- **Static Pages**: Content was served directly from the server's file system. Pages were built using

  Server-Side Includes (SSI) or Common Gateway Interface (CGI) scripting.
- **Limited Interaction**: The user experience was passive. Web 1.0 allowed for basic navigation through hyperlinks, but there was little opportunity for interaction or content creation.
- **Basic Design**: Frames and tables were used to position and align elements on pages, resulting in a rigid layout with limited flexibility.

Limitations of Web 1.0:

- **Machine Compatibility**: Web 1.0 content was designed for human readers and was not compatible with machine processing. This made it difficult for automated systems to interpret or interact with web pages.
- **Content Management**: The responsibility for updating and managing content rested solely with the website's webmaster. There was little user-generated content or community-driven updates.
- **Lack of Dynamic Representation**: Web 1.0 didn't have the ability to generate dynamic, realtime content. Users could only retrieve static information from websites.

Despite these limitations, Web 1.0 served as the foundation of the internet we know today, providing the infrastructure for digital communication, information sharing, and early e-commerce.

Web 2.0: The Rise of Interactivity and Collaboration

As the internet continued to grow and technology advanced, the shift from Web 1.0 to Web 2.0 brought about a revolutionary change. This second generation of the web was defined by its focus on user interaction, collaboration, and content sharing. The term "Web 2.0" was coined in 2004 by Dale Dougherty, but it was Tim O'Reilly who brought it into widespread use after a series of conferences discussing the internet's evolution.

In his definition, O'Reilly explained that Web 2.0 was a "business revolution in the computer industry caused by the move to the internet as a platform." The shift was not just technical—it was also cultural, as the internet became a space for collaboration, community-building, and participatory experiences.

Characteristics of Web 2.0:

- **Interactive Platforms**: Web 2.0 enabled the creation of interactive platforms where users could not only consume content but also contribute to and modify it. Social media, blogs, and forums became mainstream, allowing people to engage in online conversations and share their perspectives.
- **Social Networking**: The rise of social networking sites like Facebook, Twitter, and LinkedIn created spaces for people to connect with friends, colleagues, and like-minded individuals.
- **User-Generated Content**: Websites like YouTube, Wikipedia, and Flickr allowed users to upload and share their own content, creating a more dynamic and diverse digital ecosystem.
- **Collaborative Tools**: The introduction of tools for collaboration, such as Google Docs, wikis, and cloud-based file storage, enabled users to work together in real-time, no matter where they were located.
- **Rich Media**: Web 2.0 also saw the rise of rich media content such as video, podcasts, and interactive applications. This led to more immersive, engaging online experiences.

The Business and Cultural Impact of Web 2.0

Tim O'Reilly's definition highlighted the importance of network effects—the idea that platforms will get better; the more people use them. For businesses, Web 2.0 meant adopting a model where success was built on user engagement and community-driven content. Companies started to harness the power of usergenerated content, reviews, and recommendations to drive their business strategies.

Web 2.0 also fostered the growth of the "wisdom web," where collective intelligence could be gathered and shared among large groups of people. This made the internet a more social and participatory space, shifting the balance of power from corporations to individuals.

Web 2.0 Statistics:

- 1996: The web had 250,000 sites and 45 million global users.
- 2006: Web 2.0 saw an explosion in growth with 80 million sites and 1 billion global users, marking a significant transformation in the digital landscape.

Web 2.0 and the Shift to Collaboration

Web 2.0 didn't just introduce new technologies; it reshaped the way people interacted with the web. The emphasis on collaboration and social interaction allowed for the rise of massive online communities, each with its own set of norms, practices, and content creation tools. This era saw the democratization of the internet—more people could participate, create, and share. As we delve deeper into this transformation, it's essential to understand the major shifts that Web 2.0 introduced to the online landscape.

1. **Technology-Centric Definition**: With Web 2.0, the internet evolved into a platform that allows software applications to function across various devices. Technologies such as blogs, wikis, podcasts, and RSS feeds have empowered individuals and businesses alike to share and disseminate information easily. This shift has allowed for a more interconnected world, where users can access and interact with content seamlessly across multiple platforms.
2. **Business-Centric Definition**: Web 2.0 also revolutionized the way businesses operated online. The idea of using the internet as a platform led to the development of new business models, shifting the focus to user engagement and data-driven decisions. Businesses began leveraging the vast networks of the web to connect with customers, crowdsource ideas, and build their brands. The move towards a "networked economy" emphasized collaboration, customer interaction, and scalability.

3. **User-Centric Definition**: Perhaps the most significant aspect of Web 2.0 is the shift towards user-generated content and social interaction. This is where the "Social Web" comes into play, where users actively participate, share, and collaborate. Websites became more dynamic, encouraging user-generated content such as blog posts, reviews, comments, and more. Social networks, such as Facebook and Twitter, allow for easy communication and sharing, transforming the way people interact with one another and with brands. Platforms allow users to not just consume but also to create content and shape online communities.

4. **Dynamic and Interactive Content**: Web 2.0 brought us the era of dynamic content. Unlike the static pages of Web 1.0, the internet in Web 2.0 is more responsive. Websites can adapt to user input, making the experience more personalized and engaging. This dynamic interaction encourages users to spend more time online, enhancing their experience while enabling businesses to better target their audience.

5. **Collective Knowledge Production**: One of the hallmarks of Web 2.0 is the idea of collective intelligence. Websites like Wikipedia allow users to collaboratively create and edit content, and social media platforms provide a space for crowdsourced opinions and information. This collective knowledge has become an essential part of how we share and learn on the internet today. By tapping into the wisdom of the crowd, Web 2.0 has made information more accessible, democratic, and comprehensive.

6. **The Rise of APIs and Open Platforms**: With the development of open APIs, Web 2.0 has empowered developers to create applications that interact with existing platforms. This has facilitated the growth of a wide array of services, applications, and integrations, making it easier for developers to create innovative solutions. Through the use of APIs, websites can

now share data, offer services, and provide new functionalities, creating a more interconnected online ecosystem.

7. **Wider User Base**: Web 2.0 expanded the internet's user base far beyond its traditional techsavvy demographic. As web technologies became more accessible and user-friendly, people from all walks of life began to participate online. The shift to a more inclusive web led to a diverse range of voices and perspectives shaping the digital world. It's no longer just tech enthusiasts driving innovation; it's a broader, more diverse population.

With Web 2.0, the internet has become more dynamic and user-centric, moving beyond simple static pages to interactive and collaborative environments. Web 2.0 gave birth to a new era of digital communication, fueled by tools like:

- **Podcasting**: A medium allowing users to create and share audio content.
- **Blogging**: Platforms enabling individuals to publish written content and engage with an audience.
- **Tagging**: Organizing content through the use of metadata to make it easier to search and categorize.
- **Curating with RSS**: The aggregation of content from multiple sources, creating personalized feeds.
- **Social Bookmarking**: A method for users to save and share links to websites and resources.
- **Social Networking**: Connecting people and communities, fostering online interactions.
- **Social Media**: Platforms like Facebook, Instagram, and Twitter, where users share personal experiences, images, and opinions.

- **Web Content Voting**: Tools like Reddit or upvoting/downvoting that empower users to curate content.
- **Peer-to-Peer Payment Systems**: Technologies that enable direct transactions between users, such as PayPal or Venmo.
- **Internet of Things (IoT)**: The integration of smart devices that communicate with each other through the internet, creating a more connected world.

Challenges and Opportunities Ahead

While Web 2.0 has brought tremendous opportunities, it has also created some challenges, including issues related to privacy, security, and the centralization of power in a few tech giants. As we move into Web 3.0, we have the chance to address some of these issues. Web 3.0 promises a more decentralized, usercentric internet, offering opportunities for individuals to have more control over their data and their online experiences.

Ultimately, the evolution of the internet has been marked by increasing interactivity, user participation, and collaboration. With Web 2.0, we saw the emergence of platforms that allow people to share their voices, ideas, and creations. As we look ahead to Web 3.0, the possibilities are even greater. As we explore the evolution of the internet, the transition from Web 2.0 to Web 3.0 marks a fundamental shift in how we interact with the digital world.

Looking Ahead: The Future of Web 3.0

Web 3.0, or the "Executable Web", represents the next phase in the internet's evolution, a decentralized, personalized, and smarter web. John Markoff of The New York Times first coined the term "Web 3.0" in 2006, signifying the advent of a more intelligent, self-sustaining web. This generation of the web is driven by blockchain technology and aims to redefine how we interact with data, applications, and the internet at large. It is a web that emphasizes personalization, automation, integration, and data sharing across different platforms.

At the core of Web 3.0 lies blockchain technology, which ensures decentralized data storage, providing greater privacy, security, and transparency. This technology eliminates the need for intermediaries, allowing users to have control over their data and transactions. Web 3.0 supports the concept of a "semantic web," where machines can interpret and understand data in a more human-like way, facilitating better automation and personalization.

Some of the defining features of Web 3.0 include:

- **Decentralization**: The traditional model of centralized servers is replaced by distributed systems, making data more secure and accessible.
- **Smart Contracts**: Self-executing contracts where the terms are directly written into lines of code, automatically triggering actions when conditions are met.
- **Personalization**: Web 3.0 offers tailored user experiences by leveraging machine learning and artificial intelligence to understand individual preferences and behaviors.
- **Contextual Search**: Web searches are no longer just about keywords; Web 3.0 enables intelligent and context-driven results, leading to more relevant and personalized outcomes.
- **Interoperability**: Web 3.0 allows seamless integration across multiple platforms and applications, creating a more fluid digital ecosystem.
- **Ownership of Data**: Users have control over their personal data, allowing them to decide what to share and with whom, thus protecting privacy.

One of the most exciting aspects of Web 3.0 is the concept of blockchain-powered applications, which are poised to revolutionize industries ranging from finance to supply chain management to healthcare. These applications allow for secure, transparent, and decentralized interactions that empower users while cutting out middlemen.

To further illustrate the significance of Web 3.0, consider the disappearance of traditional webpages. In the future, instead of just viewing static pages, users will interact with data in a more immersive way, thanks to Virtual Reality (VR) and Augmented Reality (AR). These technologies will transform the internet into a dynamic and engaging experience, where content is fluid and can be personalized in real-time.

Envision an internet where individuals have more control, where innovation is decentralized, and where data and applications can interact in a smarter, more intuitive way. Web 3.0 will empower users, developers, and creators alike to participate in an evolving digital ecosystem that is more inclusive and sustainable.

As we continue to witness the growth of Web 3.0 technologies, including blockchain, AI, and decentralized applications (dApps), it's clear that this is only the beginning of a new digital era. The next steps are in our hands, and the future of the web is bright with limitless potential. Let's ensure that this transformation benefits all users, allowing for a more open, transparent, and connected digital future.

As Web 3.0 continues to unfold, it marks a significant shift in how we interact with the digital world. The Semantic Web, one of the core features of Web 3.0, is a critical evolution aimed at enabling machines to understand and process content with greater sophistication. This transition focuses on improving the way information is connected and utilized, enhancing search capabilities, and driving a more intuitive, humanlike interaction between users and data. Instead of merely relying on keywords or numeric data, the Semantic Web aims to recognize the meaning behind words and context, opening the door to a more dynamic and meaningful online experience.

Here are seven defining features of Web 3.0:

1. **Semantic Web**: The Semantic Web is the heart of Web 3.0, built upon the idea that the web should be capable of understanding and interpreting the data it holds. This allows search engines and applications to go beyond just identifying keywords, incorporating the meaning behind the words to deliver more accurate and relevant results. By recognizing relationships between concepts, the

Semantic Web aims to provide more useful and personalized experiences for users.

2. **Blockchain Technology**: At the core of Web 3.0 lies blockchain technology, which ensures secure, decentralized data storage and management. Blockchain enables transparent and immutable records, empowering users to interact with digital content, exchange assets, and perform transactions without relying on central authorities. This decentralization is key to the vision of Web 3.0, where control is distributed among users rather than monopolized by large corporations.

3. **Decentralization:** Web 3.0 is designed to decentralize the internet. Instead of relying on a few central servers or organizations, it promotes a distributed network where data and applications are stored across multiple devices and platforms. This decentralization fosters transparency, security, and user control, ensuring that individuals own and control their data rather than large corporations.

4. **Personalization:** Web 3.0 takes personalization to the next level by analyzing users' behavior and preferences to tailor content and experiences. Unlike Web 2.0, where personalization was limited to surface-level recommendations, Web 3.0 integrates artificial

intelligence and machine learning to deliver highly relevant, individualized experiences. This allows users to interact with the web in a way that feels truly unique to them.

5. **Artificial Intelligence (AI) and Machine Learning:** Artificial Intelligence and machine learning are integral components of Web 3.0, enabling websites and applications to understand and learn from user interactions. These technologies enhance search engines, virtual assistants, and content recommendations, helping to create smarter and more responsive digital environments. AI's role in Web 3.0 goes beyond automation; it aims to make the internet more intuitive, context-aware, and capable of predicting and serving user needs.

6. **Interoperability:** Web 3.0 strives for interoperability, meaning that various applications, platforms, and systems can communicate with one another seamlessly. This is made possible by blockchain technology and standardized protocols, which ensure that users can access, share, and interact with data and services across multiple platforms without friction. This interoperability fosters a more connected and integrated web, where digital assets and information can flow freely between different environments.

7. **Privacy and Security:** Privacy and security are fundamental to the evolution of Web 3.0. As more data is shared and transferred across decentralized networks, ensuring that users' personal information remains safe and under their control is paramount. Blockchain technology plays a pivotal role in enhancing security by offering cryptographic protection and enabling secure transactions without relying on centralized entities. Web 3.0 also emphasizes transparency, allowing users to track and verify their data usage and movements online.

Web 3.0: Gaming with Virtual and Augmented Reality

Web 3.0 has fundamentally altered the way humans interact with digital technology, particularly through the rise of two major concepts: Virtual Reality (VR) and Gaming. The gaming industry, in particular, has experienced tremendous growth, with online games offering boundless opportunities for creativity, exploration, and interaction. Unlike physical games, which are often limited by space and resources, web-based gaming allows for a more expansive and immersive experience. This shift in gaming trends signals that online platforms are likely to overtake traditional physical games in the near future.

As Web 3.0 has grown, it has paved the way for further innovations in Virtual Reality (VR) and Augmented Reality (AR) technologies. Gamers can now experience more dynamic, participatory, and engaging environments, where their interactions feel more real and synchronized with the virtual world. The introduction of VR headsets and AR devices marks the beginning of a new gaming era, where players are no longer observers but active participants in an evolving, interactive universe. Web 3.0 is no longer just about digital advancement—it's about creating a new reality where gaming, entertainment, and creativity merge seamlessly.

Looking Ahead to Web 4.0: Taking the Immersive Experience Even Further

Born from the rapid evolution of Blockchain technology and Artificial Intelligence (AI), Web 4.0 will usher in an era where digital interactions become more intelligent, intuitive, and seamless. AI-powered systems, such as smart chatbots and intelligent assistants, will become common tools that help users search for and access information in real-time. These systems will be able to respond quickly and more accurately to user needs, creating a far more efficient and personalized internet experience.

Web 4.0 will also include new dimensions like 3D animation and voice interaction, which will enhance communication between humans and machines. Mobile devices will play a crucial role in this evolution, allowing users to access and share information from anywhere, without the need for a desktop computer. The convenience and mobility of Web 4.0 will extend the internet reach far beyond what we currently experience, making it accessible and adaptable to anyone, anywhere.

However, the exponential growth of the Internet of Things (IoT), with an estimated 200 billion connected devices by 2025-2026, raises some important concerns. As more devices communicate with each other, AI systems will have the capability to track, monitor, and analyze human behavior in real-time. While this offers unparalleled convenience and data collection opportunities, it also presents significant challenges to privacy. As data from our daily interactions is continuously monitored and controlled, the question arises: how much of our personal data will we be willing to share in the name of convenience?

The integration of AI with IoT also enables direct communication between devices across vast distances, creating a web of interconnected networks. This could ultimately lead to the creation of The Metaverse, a fully immersive virtual world where people can interact, work, and live in parallel with our physical existence. Through VR headsets and advanced AI systems, users will be able to participate in this virtual space, offering unprecedented opportunities for collaboration, entertainment, and digital interaction.

At the core of Web 4.0 lies a major shift in how we interact with technology. The lines between humans and machines will blur, creating what can be referred to as a symbiotic relationship, similar to the cyborgs in *Star Trek*. As technology advances, humans will have constant access to robots, chatbots, and AI, which will become integrated into our everyday lives. This cyborg-like existence, where humans and machines work together in harmony, will redefine the very essence of human experience in the digital age.

With all these possibilities come significant ethical questions. As technology evolves, how do we ensure that it remains beneficial and not harmful? Will the conveniences of Web 4.0 come at the cost of our personal freedoms, privacy, and security? The future of the internet holds immense promise, but we must be cautious in shaping a future that respects individual rights while embracing technological advancements.

The next era of the World Wide Web is approaching, and the power to shape it is in our hands.

Web 4.0 is set to revolutionize the way we interact with the digital world, creating a closer bond between humans and artificial intelligence. Here are some of the key features of the Web 4.0 period:

1. **Symbiotic Relationship Between Man and AI:** Web 4.0 ushers in an era where the line between humans and machines is increasingly blurred. The relationship is no longer just user and tool, but a partnership where artificial intelligence learns from, adapts to, and enhances human capabilities. This interaction leads to a truly symbiotic relationship between man and AI, fundamentally changing how we live and work.

2. **The Internet Becomes a Web OS**: The concept of Web 4.0 transforms the internet into a comprehensive operating system (OS) where all information can be shared seamlessly across the globe. It connects devices, services, and data points, making information access faster, more efficient, and universally available.

3. **Active Use of Artificial Intelligence**: Artificial intelligence is no longer a passive tool but an active participant on the web. AI algorithms analyze, interpret, and predict user behavior, enhance experiences, and offer personalized services. From chatbots that converse like humans to AI-driven content recommendations, the web will actively improve its services in real-time based on the user's preferences and actions.

4. **Human-like Communication**: Web 4.0 aims for a level of interaction where the internet communicates with users in the same way humans do. Think of AI chatbots that can hold natural conversations, or virtual assistants that understand and respond to nuanced commands. This humanlike interaction will help bridge the gap between technology and users, creating a more intuitive and engaging web experience.

5. **Transparency and Traceability**: In the Web 4.0 era, every transaction and action will be traceable, thanks to the decentralized nature of blockchain and the advancement of AI analytics. While this promises to increase security and accountability, it also raises privacy concerns as every user action could potentially be tracked.

6. **Speed and Reliability**: With advancements in infrastructure, Web 4.0 promises faster internet speeds and more reliable connections. The web will not only offer greater capacity for information exchange but also ensure that users can rely on it for real-time actions with minimal latency.

Web 4.0's Impact on Daily Life

Web 4.0 doesn't just bring technological advancements—it changes the very fabric of everyday. life From business operations to healthcare, logistics, and even entertainment, Web 4.0's transformative nature offers a deeper connection between people and machines. It gives consumers more control over their personal data, offering them greater autonomy in the digital space.

However, such advancements come with potential risks. The rise of super-intelligent AI has sparked concerns over its possible implications. Could AI surpass human intelligence and act independently? Elon Musk has repeatedly raised alarms about AI's growing capabilities, highlighting the need for humans to keep pace with this technology.

This is one of the reasons Musk has been working on Neuralink, a neurotechnology company developing brain-machine interfaces (BMIs). Neuralink's goal is to enhance human cognitive abilities by creating devices that connect the brain to machines. While currently focused on therapeutic uses, the implications of such technology could extend far beyond medical applications. It could enable humans to directly interface with machines, essentially creating a cyborg-like relationship.

The Terminator Effect

As we move toward the integration of humans and AI, it's hard to ignore the dystopian themes that often arise in science fiction, particularly in movies like Terminator. With AI becoming increasingly smarter, the question arises: Could machines one day act independently and make decisions that may not align with human interests? While AI promises to augment human capabilities, we must remain vigilant about the potential dangers of AI evolving beyond our control. The fear of AI making decisions for us, or even against us, is something we cannot ignore as we step into this new era of Web 4.0.

In summary, Web 4.0 presents both incredible opportunities and serious challenges. As we integrate AI, blockchain, and other advanced technologies into every aspect of our lives, we must carefully consider the ethical and security implications of this new digital frontier. The future of the internet is both exciting and daunting, and it is up to us to ensure that technology serves humanity rather than becoming a force that we can no longer control.

As we stand on the precipice of technological advancements that are rapidly transforming how we live, it's essential to remember the ethical considerations and the implications of these developments on our society. The integration of artificial intelligence, blockchain technology, and neurotechnology like Neuralink is undeniable. These innovations bring incredible benefits, but they also raise serious questions about privacy, autonomy, and human identity.

The allure of enhanced cognitive capabilities or more efficient ways to interact with technology is compelling, but it also presents risks. The possibility of machines taking control over sensitive data, influencing decisions, or even manipulating our actions is something that cannot be ignored. As much as technology can improve our lives, we must ensure it doesn't strip away our fundamental freedoms.

The power of choice remains in our hands. It’s crucial that we stay informed, think critically, and use technology responsibly. While the idea of human-machine integration might seem fascinating or futuristic, it's important to reflect upon the values that define our humanity—our ability to choose, to think freely, and to live with dignity.

In the face of these advancements, we need to strike a balance. It's okay to explore and embrace new technologies, but we must not forget the core principles that guide us: the value of human life, the importance of privacy, and the need for ethical conduct. This is why it's essential to stay grounded, do your own research (DYOR), and remain true to what feels right for you.

At the end of the day, we are the stewards of our future. The choices we make today will shape the world tomorrow, and it’s crucial to ask ourselves: What kind of world do we want to live in? A world where we retain control of our decisions, or one where machines dictate our every move? The answer lies within each of us.

As we move forward, let us remember that technology should serve humanity, not the other way around. Our humanity is too valuable to lose in the pursuit of progress.

Chapter 15

DAOS: The New Autonomous Internet Community

Decentralized Autonomous Organizations, or DAOS, are community-owned entities that operate on a blockchain with no central leadership. The rules are enforced automatically, and the community governs everything from how decisions are made to how funds are allocated. The idea is that there's no need for intermediaries, like board members or executives, to manage the organization or make decisions. Instead, all stakeholders, or token holders, have an equal say in the operation of the DAO.

How Does a DAO Work?

A DAO operates through smart contracts on a blockchain, which are self-executing contracts with the terms of the agreement directly written into code. These contracts are immutable, meaning once they are set, they cannot be changed unless agreed upon by the majority of participants.

Here's how the process generally works:

1. **Smart Contracts and Governance**: The smart contracts define the rules of the DAO. These rules might include things like how funds should be spent, who has the right to vote on decisions, and what processes need to be followed for different actions. These rules are transparent, publicly visible, and enforced automatically without human intervention.
2. **Tokens and Voting**: In a DAO, tokens are used as voting power. Stakeholders hold these tokens, and their votes are weighted according to the number of tokens they own. This voting system allows the community to make collective decisions about the future direction of the project or organization. The more tokens you hold, the greater influence you have, but every holder has an equal opportunity to participate.

3. **Decentralization:** One of the most important aspects of a DAO is decentralization. Unlike traditional corporations where a select group of executives and directors make decisions on behalf of the shareholders, DAOs allow stakeholders to have a direct say in the decision-making process. This eliminates the need for a centralized leadership structure and ensures that the community, rather than a few individuals, controls the direction of the organization.

Why Are DAOs Important?

DAOs are revolutionary because they provide a new form of governance, allowing organizations to operate more transparently, efficiently, and equitably. The key benefits include:

- **Decentralization**: DAOs remove the need for central authority, which can lead to more democratized and fair decision-making.
- **Transparency**: All transactions and decisions within a DAO are recorded on the blockchain, meaning they are visible to all participants. This promotes trust and accountability.
- **Global Participation**: Because DAOs are decentralized and run on blockchain technology, anyone, anywhere in the world, can participate. This opens up the opportunity for global collaboration without the need for geographical boundaries.
- **Autonomy**: Smart contracts execute actions automatically based on pre-defined rules. This removes the potential for human error, corruption, or manipulation in the decision-making process.

Examples of DAOs in Action

Several high-profile projects and platforms in the cryptocurrency space have embraced the DAO model. Here are some examples:

1. **MakerDAO**: MakerDAO governs the Maker Protocol, which allows users to create and manage a stablecoin called DAI. MakerDAO is governed by MKR token holders, who vote on decisions like collateral types, stability fees, and other parameters of the system. The entire protocol operates without a central authority, and decisions are made by the community.
2. **Uniswap**: Uniswap, a decentralized exchange (DEX), transitioned to a DAO structure with the introduction of its UNI token. UNI holders have voting power to influence the direction of the protocol, such as deciding on fee structures or the allocation of treasury funds.
3. **Aragon**: Aragon provides a platform for creating and managing DAOs. It enables users to build and govern decentralized organizations without needing technical expertise. Aragon makes it easier for communities to create their own DAOs and manage them with governance tools built on the Ethereum blockchain.
4. **Aave**: Aave is a decentralized lending and borrowing platform that has incorporated DAO governance through its AAVE token. Token holders vote on proposals related to the protocol development, such as adding new assets to the platform or modifying lending rates.

The Future of DAOs

DAOs represent a radical shift in how organizations and businesses can function. By removing centralized intermediaries and offering more direct control to stakeholders, DAOs are poised to disrupt industries ranging from finance to governance and even supply chain management. They enable truly decentralized projects to flourish and have the potential to revolutionize how decisions are made in many sectors of the economy.

However, it’s important to remember that DAOs are still in their early stages, and many challenges remain. Issues such as governance disputes, scalability, and the legal recognition of DAOs are still being explored. As the technology matures, we can expect to see more widespread adoption and refinement of DAO models.

DAOs are an exciting and innovative concept that exemplifies the true potential of blockchain technology. They empower individuals, foster decentralization, and offer a more democratic approach to decision-making. If you're new to the concept, now is the perfect time to dive deeper and explore how DAOs might fit into your understanding of the future of the internet and decentralized communities.

While DAOs represent one of the most powerful expressions of decentralization, their evolution has not been without defining moments that tested their foundations. One of the earliest and most significant events in DAO history was the rise and collapse of The DAO — an ambitious experiment built on Ethereum that aimed to operate as a fully decentralized venture fund.

The DAO attracted massive attention and participation, signaling strong belief in the model. However, a vulnerability within its smart contract was exploited, allowing an attacker to drain funds through a recursive withdrawal mechanism embedded in the code. The issue was not a failure of blockchain itself, but a flaw in the contract design — a critical oversight that went undetected before deployment.

This incident became a turning point for the entire ecosystem. It forced the Ethereum community to confront complex questions about security, governance, accountability, and the meaning of immutability. The response to this crisis would ultimately reshape Ethereum's future and redefine how decentralized systems approach risk and responsibility.

The Aftermath of the DAO Hack

The hack on The DAO was a pivotal moment in the history of blockchain technology, as it demonstrated both the potential and the risks of decentralized applications and smart contracts. The breach created a crisis of confidence in the Ethereum network and raised serious concerns about the security of decentralized systems.

In response to the hack, the Ethereum community had to make a critical decision: how to handle the stolen funds. One possible solution was to let the hacker keep the stolen Ethereum, which would have led to a loss of trust in the network. The other option was to "fork" the blockchain to reverse the hack and return the stolen funds to the original investors.

The Ethereum Hard Fork

After intense debate and discussions, the Ethereum community opted for a hard fork to restore the stolen Ethereum to its original owners. This decision was not unanimous, and it created a significant divide within the community. Those who opposed the fork believed that the blockchain should be immutable, meaning that transactions should not be altered or reversed under any circumstances. They argued that reversing the hack would undermine the core principle of decentralization and the immutability of the blockchain.

Despite the opposition, the majority of the Ethereum community supported the hard fork, and in July 2016, Ethereum underwent a contentious fork, creating two separate blockchains: Ethereum (ETH) and Ethereum Classic (ETC). Ethereum, the new chain, reversed the effects of the hack and returned the stolen funds to the investors. Ethereum Classic, the original chain, maintained the transaction history as is, allowing the hack to stand.

The hard fork — and the resulting split within the Ethereum community — exposed a fundamental tension at the heart of decentralization: the balance between immutability and fairness. One side believed the blockchain should remain untouched under any circumstances, preserving the principle that transactions are irreversible. The other argued that intervention was necessary to protect participants and restore trust.

The decision ultimately led to two paths. Ethereum moved forward with the fork and continued expanding in value, development, and adoption, while Ethereum Classic remained on the original chain, supported by a smaller but deeply committed community that upheld immutability above all else.

What was once envisioned as a revolutionary leap in decentralized investment and governance had been shaken by a single vulnerability. The exploit served as a powerful reminder that the strength of decentralized systems depends entirely on the integrity of their code. Smart contracts, once deployed, operate autonomously — and any flaw can carry irreversible consequences.

Although the incident was a major setback for The DAO and a defining crisis for Ethereum, it also, sparked critical conversations about security standards, risk management, and the future of decentralized finance and governance. In many ways, the lessons learned from this moment laid the groundwork for the more resilient DeFi ecosystem that would emerge in the years that followed.

The Legacy of the DAO

Although The DAO itself was never fully recovered after the hack, the lessons learned from its creation and failure have had a lasting impact on the development of decentralized finance (DeFi) and blockchain governance.

1. **Smart Contract Auditing and Security**: The DAO hack underscored the importance of thorough auditing and testing for smart contracts before deployment. It led to the development of more sophisticated security practices and tools for detecting vulnerabilities in smart

contract code. As a result, security audits have become a standard part of the development process for DeFi projects today.

2. **Blockchain Governance**: The DAO's decentralized governance model, in which token holders could vote on proposals and projects, sparked interest in creating new types of decentralized organizations. Although the DAO itself failed, its idea of community-led decision-making through token-based voting inspired the creation of other DAOs and the growth of decentralized governance systems in the blockchain ecosystem.

3. **Crowdfunding on the Blockchain**: The DAO's success in raising $150 million demonstrated the potential for blockchain to disrupt traditional crowdfunding and venture capital funding models. It showed that, with the right incentives, large-scale funding could be raised without relying on traditional intermediaries. Since then, many other blockchain-based crowdfunding and venture capital projects have emerged, using DAO-like structures to raise funds and make decisions collectively.

4. **Ethereum's Growth**: Despite the controversy surrounding the hard fork, Ethereum continued to grow and evolve. The decision to perform the hard fork ultimately strengthened Ethereum's reputation as a blockchain capable of making decisions and resolving issues quickly. Ethereum also became the platform of choice for many other decentralized applications, including those in the DeFi space.

5. **Ethical and Legal Implications**: The DAO hack and subsequent hard fork raised important questions about the ethical and legal implications of decentralized systems. Should blockchain protocols be allowed to alter transactions in response to unforeseen circumstances? What happens when the decentralized community makes a decision that not everyone agrees with?

These questions remain at the forefront of discussions about the future of decentralized governance and blockchain technology.

The Rise of DAOs Today

Although The DAO itself was a failure, it played an essential role in shaping the future of decentralized organizations. Since then, the DAO model has been adopted by various projects in the cryptocurrency and blockchain space. Today, DAOs are used to manage everything from DeFi protocols to NFT platforms, with a wide range of voting mechanisms, governance models, and tokenomics.

Some notable DAOs today include:

- **MakerDAO**: Governs the Maker Protocol, which powers the stablecoin DAI. MakerDAO token holders vote on changes to the protocol, such as which assets are used as collateral and the interest rates for borrowing.
- **Uniswap**: A decentralized exchange (DEX) that transitioned to a DAO structure, allowing UNI token holders to vote on proposals related to protocol upgrades and the distribution of funds from the governance treasury.
- **Aave**: A decentralized lending platform that uses a DAO to govern decisions related to the protocol's development and operations, with token holders voting on key proposals.
- **Aragon**: A platform for creating and managing DAOs that allows users to create decentralized organizations without needing technical expertise.

The DAO and the events surrounding it marked a crucial chapter in the history of blockchain technology and decentralized governance. While it faced challenges, including a high-profile hack, the DAO experiment provided valuable insights that have shaped the development of DeFi and DAO governance systems today. In the future, DAOs will likely continue to play an increasingly important role in how decentralized communities, organizations, and businesses operate, making blockchain governance and participation more accessible and transparent for everyone.

Fast forward to 2020 and 2021, and DAOs have evolved significantly. One notable example is MakerDAO, the creator of the DAI stablecoin, which made headlines when it officially transitioned its operations to a decentralized governance model. In 2021, the Maker Foundation, a pivotal entity in the development of DAI, announced that it would dissolve and transfer full control of the Maker protocol to the community. This marked a key milestone in the rise of DAOs as viable, self-sustaining entities that operate independently of centralized foundations.

The MakerDAO transition is a testament to the growing confidence in DAOs and their ability to govern decentralized platforms. By giving the community control over key decisions, MakerDAO created a system that is not only transparent but also resilient to external pressures. The MakerDAO community, composed of token holders, governs the project by participating in proposals, voting, and discussions regarding the protocol's future. This participatory model ensures that the voice of the community is heard, empowering individuals to shape the direction of the project.

The growing success of DAOs has also brought forth numerous innovations. For example, some DAOs are now experimenting with decentralized finance (DeFi) protocols, governance tokens, and even decentralized insurance platforms. The beauty of DAOs lies in their adaptability—whether they are used for community decision-making, decentralized lending, or governance over autonomous organizations, DAOs enable users to take part in the future of the internet and finance in a way that was never possible before.

Examples of successful DAOs today include:

- MakerDAO: As mentioned earlier, MakerDAO governs the DAI stablecoin, with community members voting on important decisions like collateral types, risk management, and stability fees.

- Uniswap: A decentralized exchange (DEX) governed by its community through the UNI token.

  Token holders vote on proposals related to the protocol's development and future direction.

- Aave: A decentralized lending and borrowing platform where the AAVE token holders participate in governance and protocol upgrades.

As the world increasingly turns to decentralized platforms, DAOs have become a driving force in the evolution of blockchain technology. They are reshaping industries such as finance, governance, gaming, and social organizations, bringing a new level of transparency and community involvement to the forefront. The continued growth of DAOs suggests a bright future where individuals have a direct say in the way organizations are run, and the influence of centralized entities continues to decrease.

The benefits of DAOs are clear: they enable more equitable decision-making, provide transparent governance structures, and reduce reliance on intermediaries. By allowing token holders to vote and participate in critical decisions, DAOs offer a decentralized, community-driven approach to governance that could have profound implications for everything from financial systems to the future of global collaboration.

With the lessons learned from The DAO hack, the rise of successful projects like MakerDAO, and the continued exploration of decentralized governance, it's evident that DAOs have evolved into a robust, secure, and promising model for the future of blockchain technology and beyond. As more projects adopt

DAOs as their governing structure, we are witnessing the dawning of a new era in the way organizations are structured and managed—one that prioritizes community input, transparency, and security.

DAO	**A traditional organization**
Usually flat, and fully controlled democratized.	**Usually hierarchical.**
Voting power required by members for any changes to be implemented.	**Depending on structure, changes can be demanded from a sole party, or voting may be offered.**
Votes are tallied, and outcome implemented automatically without trusted intermediary.	**If voting allowed, votes are tallied internally, and outcome of voting must be handled manually.**
Services offered are handled automatically in a decentralized and organized manner (for example distribution of philanthropic funds).	**Requires human handling, or centrally controlled automation, prone to manipulation.**
All activity is transparent and fully public.	**Activity is typically private, and limited to the public.**

Examples of How DAOs Are Used Today

To further illustrate how DAOs are being used today, here are a few examples of how you could leverage a DAO for various purposes:

- **A Charity**: A DAO can be established to accept membership and donations from anyone globally. Members can vote on how the collected funds should be spent, ensuring the community has a say in the allocation of resources.
- **A Freelancer Network**: Freelancers can pool their funds together to share resources, such as office spaces or software subscriptions. Members of the DAO can vote on how these shared resources should be managed and distributed.
- **Ventures and Grants**: A venture fund could be set up as a DAO, where investors pool their capital and vote on which ventures or projects to back. Any returns or profits could later be redistributed among DAO members based on their token holdings.
- **Startup**: A DAO could be used to help govern a startup business. The community could provide feedback, suggest ideas, and make decisions collectively, giving them a direct influence over the company's future.

Types of DAO Memberships

DAOs often offer different types of memberships, each influencing how voting rights and other key

features operate within the organization. Below are the two main types of memberships you might encounter:

1. **Token-Based Membership**:

- These DAOs are usually fully permissioned, depending on the tokens used. Governance tokens can often be traded on decentralized exchanges. Some tokens must be earned through actions like providing liquidity or completing a 'proof-of-work' task. Simply holding a certain amount of tokens grants the holder voting rights within the DAO.
- This membership is typically used to govern decentralized protocols or tokens themselves.

2. **Share-Based Membership**:

- Share-based DAOs offer a more permissioned approach but are still quite open. New members can submit proposals to join the DAO by offering value in the form of tokens or work. Shares represent direct voting power and ownership in the organization. Members can leave at any time and take their proportionate share of the DAO's treasury with them.
- These types of DAOs are often used for organizations like charities, worker collectives, or investment clubs, and can also govern tokens and protocols.

Top DAOs by Market Capitalization

The DAO ecosystem has grown significantly over the years. As of today, the total market cap of DAOs stands at $25,071,586,842, with over 140 different DAOs to choose from. Here is a list of some of the top DAO tokens by market capitalization:

- Uniswap
- Alpha DAO
- Lido DAO

- Apecoin
- Aave
- Maker
- Dash
- Curve DAO Token
- Compound
- Decred
- Kyber Network Crystal V2
- Synthetix
- UMA
- Ox
- SushiSwap
- Illuvium
- Nervos Network
- Ethereum Name Service
- BitDAO
- Aragon
- Rari Governance Token
- Alien Worlds

- Cream Finance
- Badger DAO
- Mantra DAO

Remember, as with any investment, always perform your own research (DYOR) before participating in any DAO. These projects are highly innovative, but also carry risks, so ensure that you fully understand the functionality and goals of a DAO before getting involved.

The shared examples highlight the success of various DAOs in the market, with Apecoin leading the way as a recent and impactful player. Apecoin, associated with the well-known Bored Ape Yacht Club NFTs, launched on March 17, 2022, and has quickly climbed to a ranking of 35th in market capitalization, with a 24-hour trading volume of $8.35 billion. This surge is a testament to the growing trust and adoption of DAOs in the market. The success of these decentralized organizations emphasizes the powerful connection that DAOs create between communities and developers.

Looking ahead, the future of DAOs is bright as they continue to redefine how communities and businesses interact. With the ability for users to directly participate in decision-making processes, DAOs enable projects to evolve based on collective input. This has created a stronger bond between developers, corporations, and users, further solidifying the role of DAOs in shaping the future of the decentralized ecosystem.

Are you ready to be part of this revolutionary change? DAOs offer an opportunity for your ideas, projects, and dreams to come true with the support of a community that believes in your vision. The transparency, security, and decentralized nature of DAOs make them a great tool for building trust and fostering growth in the blockchain space.

Before diving in, it's important to DYOR (Do Your Own Research), connect with communities through platforms like Twitter and Discord, and engage with those who share your vision. The possibilities are vast, and the future is here. What are you waiting for? The world of DAOs is waiting for you!

Chapter 16

The Power of Atomic Swaps: A Game-Changer for Cryptocurrency Trading

What if I told you there's a way to swap your cryptocurrency pairs on decentralized exchanges (DEXs) while only paying network fees—and avoiding the extra charges that centralized exchanges like Coinbase often impose for deposits, withdrawals, or even trading? That's right—no more worrying about withdrawal fees, deposit fees, or those steep trading fees that cut into your profits. In this chapter, we'll explore a little-known feature of the cryptocurrency space that many have yet to fully utilize—atomic swaps. So, what exactly are atomic swaps? How do they work, and why should you care?

What Are Atomic Swaps?

Atomic swaps are a revolutionary breakthrough in cross-chain trading, enabling the seamless exchange of digital assets from one user to another, without relying on a centralized third party, such as a cryptocurrency exchange. The term "atomic" comes from the idea that these transactions are either completely successful—with both parties receiving the agreed-upon funds—or fail entirely—with both parties retaining their original assets. It's like an on/off switch, where the trade happens or doesn't, with no middle ground. This trustless, peer-to-peer approach is made possible via decentralized applications (DApps) and decentralized exchanges (DEXs), where transactions occur directly between wallets.

Atomic swaps can happen between cryptocurrencies from different blockchains—for example, between Bitcoin and Ethereum—without requiring an intermediary. These swaps work through a mechanism known as Hashed Time-Locked Contracts (HTLCs), which ensures that the trade will only take place if both parties fulfill their obligations, or else the trade is canceled and no funds are lost.

How Do Atomic Swaps Work?

Let's break down how an atomic swap works using a real-life example:
Imagine Lisa has 20 Ethereum (ETH) and she wants to swap it for Apecoin. Here's how the process unfolds:

1. **Connect Wallet:** Lisa connects her wallet (e.g., MetaMask) to a decentralized exchange, like Uniswap.
2. **Initiate Swap**: She enters the amount of Ethereum (ETH) she wants to swap for Apecoin. A popup appears on MetaMask, showing the network fee (gas fee), which is the only fee she will pay for this transaction.
3. **Transaction Signing**: Lisa reviews the details and clicks "Confirm." This signs the transaction, allowing it to be sent to the blockchain for confirmation.
4. **Blockchain Confirmation**: The transaction enters the blockchain, going through several confirmations. Once confirmed, the swap is complete, and Lisa now holds Apecoin in her wallet.

In this scenario, Lisa only paid a network fee (the gas fee) for the swap. There are no additional fees for deposits, withdrawals, or trading, which you would typically encounter on centralized exchanges like Coinbase or Binance. Furthermore, the process is streamlined, and she maintains control of her private keys, a key feature that centralized exchanges don't offer.

Why Atomic Swaps Are Better Than Centralized Exchanges

One of the most significant advantages of atomic swaps over centralized exchanges is that you retain full control of your funds. When you use a centralized exchange like Coinbase, Kraken, or Binance, you are required to deposit your funds into the exchange's wallet. This means that you lose control over your private keys, which poses a risk to the security of your funds.

If the exchange is ever hacked or faces a technical failure, you may lose your assets, and there may be limited recourse in recovering your stolen funds. Additionally, centralized exchanges have the power to freeze your account, block withdrawals, or even impose withdrawal limits, leaving you with no access to your assets.

With atomic swaps, your funds stay in your wallet throughout the entire transaction process. You are not relying on a third party to hold or manage your funds. This means that you are the sole custodian of your private keys, which enhances security and privacy. If you want to maintain complete control over your assets, decentralized exchanges powered by atomic swaps offer the most secure and cost-effective solution.

The Problem with Centralized Exchanges

Centralized exchanges have their own set of drawbacks that often frustrate traders:

- **Custodial Risk**: As mentioned earlier, when you deposit your funds on a centralized exchange, you're essentially handing over control to that exchange. If something goes wrong—whether it's a hack, a bug, or the exchange going out of business—your funds are at risk.
- **High Fees**: Centralized exchanges typically charge fees for deposits, withdrawals, and trades. These fees can eat into your profits, especially if you trade frequently. Atomic swaps, on the other hand, only incur network fees—no hidden charges or withdrawal fees.

- **Account Freezes**: Centralized exchanges have the power to freeze accounts and restrict access to your funds. This could happen if the exchange suspects fraudulent activity, or in some cases, if governments or regulatory bodies put pressure on the platform. Atomic swaps eliminate this risk by ensuring users retain full control over their assets at all times.

Security: Cold Wallets vs. Hot Wallets

When trading on centralized exchanges, many users store their assets in hot wallets, which are connected to the internet and are more vulnerable to hacks. Conversely, cold wallets (hardware wallets like Ledger or Trezor) are offline, making them far less susceptible to cyberattacks.

While cold wallets are the most secure way to store your cryptocurrencies, they're not ideal for frequent trading. However, some hardware wallets allow users to stake their assets while in cold storage, providing flexibility for those who want to hold their assets securely but still engage in DeFi activities.

With atomic swaps, you don't need to rely on a centralized exchange or keep your assets in a hot wallet. You can trade directly from your cold wallet or any secure wallet without exposing your funds to unnecessary risks.

The Evolution and History of Atomic Swaps

The early history of atomic swaps dates back to 2012 when Sergio Demian Lerner introduced the idea of a trustless, peer-to-peer cryptocurrency exchange protocol, later known as P2PTradeX. This concept would later lay the foundation for atomic swaps. However, it wasn't until 2014 that the concept became a reality when developer jl777, from the Komodo project, created the first code that allowed swapping between Litecoin and Dogecoin using atomic swap protocols. The first successful test of atomic swaps took place using the UTXO (Unspent Transaction Output) model, which laid the groundwork for the use of atomic swaps across a variety of blockchain networks.

Atomic swaps gained traction over the next few years as developers recognized the potential for creating decentralized, trustless exchange systems that eliminated the need for centralized exchanges. In 2017, Decred improved on the concept by making the process more accessible. In September 2017, Decred successfully performed an atomic swap between Litecoin and Bitcoin, effectively proving that users could trade directly with one another on different blockchain networks without relying on centralized third parties.

This marked the beginning of a new phase for atomic swap technology.

A significant milestone in the development of atomic swaps was achieved in November 2017, when the Lightning Network conducted the first off-chain atomic swap between Bitcoin and Litecoin. These offchain swaps, made possible by the Lightning Network, highlighted the potential for faster, more scalable swaps that do not require the high fees and delays typical of on-chain transactions. The introduction of the Lightning Network provided the necessary infrastructure for atomic swaps to scale beyond the initial demonstrations.

Komodo's work with the BorderDEX platform further simplified the process of conducting atomic swaps. In 2017, Komodo introduced the first graphical user interface (GUI) for automated atomic swaps, enabling cryptocurrency traders to access decentralized trading platforms more easily. Komodo continued to innovate with the release of AtomicDEX, a third-generation decentralized exchange, which allowed users to conduct atomic swaps on mobile devices. This made atomic swaps accessible to a much wider audience and marked a major step forward in blockchain interoperability.

By 2018, Komodo had successfully bridged Ethereum with Bitcoin-protocol coins, with an atomic swap between Ethereum and Dogecoin. This was a critical achievement as it showcased that atomic swaps could function across different blockchain protocols, further expanding the potential of decentralized trading. The move to include Ethereum was a crucial development in the evolution of atomic swaps, as Ethereum's smart contract capabilities significantly enhanced the technology's utility.

The continued growth of atomic swap technology culminated in the release of AtomicDEX, which allowed users to perform cross-chain swaps on a fully decentralized platform, without needing to trust a central authority. By 2020, Komodo expanded AtomicDEX to desktop devices, making it easier for users to trade assets securely and privately from their computers. This development positioned Komodo as one of the most important players in the atomic swap space, continuing to innovate and improve the technology for decentralized finance.

What Makes Atomic Swaps Possible

Now that we've covered the history of atomic swaps, it's important to understand the underlying technical features that make them possible. Two key components are essential to the functioning of atomic swaps: OP_CHECKLOCKTIMEVERIFY (CLTV) and Pay-to-Script-Hash (P2SH) transactions.

1. **OP_CHECKLOCKTIMEVERIFY (CLTV)**: Introduced through Bitcoin Improvement Proposal (BIP-65) in October 2015, the CLTV command plays a critical role in atomic swaps. CLTV ensures that a transaction cannot be spent until a predetermined time has passed. This time-lock mechanism prevents the receiver from accessing the funds prematurely, ensuring that both parties in the atomic swap transaction adhere to the terms. This feature enables trustless exchanges, as it guarantees that a payment will be released only when both parties fulfill the conditions of the swap.

2. **Pay-to-Script-Hash (P2SH)**: P2SH transactions allow users to send funds to an address that requires special conditions before the funds can be spent. With P2SH, the funds are locked within a script, and the receiver needs to fulfill certain criteria to unlock and access the funds. In the context of atomic swaps, this feature ensures that both parties in the swap must meet the conditions of the agreement before any funds are released. This is essential for ensuring that the atomic swap remains trustless and secure.

In atomic swaps, these two features work together to guarantee that the transaction will either be completed in full, with both parties exchanging their assets as agreed, or nothing will happen at all. This "all or nothing" characteristic is what makes atomic swaps "atomic"—it's a fail-safe system that ensures neither party is left at a disadvantage.

The Future of Atomic Swaps

Looking ahead, the future of atomic swaps is promising. As blockchain technology continues to evolve, atomic swaps will likely play an increasingly central role in the decentralized finance (DeFi) ecosystem. They provide a viable alternative to centralized exchanges, offering users more privacy, lower fees, and greater control over their assets. With further improvements in blockchain interoperability, atomic swaps could become even more widespread, facilitating seamless exchanges across a multitude of cryptocurrencies and decentralized networks.

In conclusion, atomic swaps have revolutionized the way we think about cryptocurrency exchanges. By eliminating the need for third-party intermediaries, they provide users with greater control, security, and efficiency. As the technology continues to develop, atomic swaps will likely become a core feature of the decentralized finance ecosystem, enabling users to exchange digital assets directly, securely, and without the heavy fees associated with centralized exchanges. Whether you're an individual cryptocurrency enthusiast or a developer looking to integrate cross-chain swaps into your project, atomic swaps offer an exciting and promising future for the cryptocurrency space.

Top Decentralized Exchanges for Atomic Swaps

For those looking to trade cryptocurrencies using atomic swaps, there are several decentralized exchanges (DEXs) that provide this feature. Here's a list of some of the top DEXs that support atomic swaps and peer-to-peer trading:

1. Uniswap – uniswap.org
2. AtomicDEX – atomicdex.io
3. SushiSwap – sushi.com
4. PancakeSwap – pancakeswap.finance
5. Phantom – phantom.app
6. SpookySwap – spookyswap.finance
7. 1Inch – 1inch.io
8. Atomic Swap Wallet – atomicswapwallet.io
9. Exodus – exodus.com
10. Jaxx Liberty – jaxx.io
11. Binance DEX – binance.com
12. AnySwap – anyswap.exchange
13. AirSwap – airswap.finance

14. Balancer – balancer.fi

15. Bancor – bancor.network

16. Curve Finance – curve.fi

17. IDEX – idex.io

18. JellySwap – jelly.market
19. KyberSwap – kyberswap.com

20. Liquality – liquality.io

21. ParaSwap – paraswap.io

22. QuickSwap – quickswap.exchange

23. Raydium – dex.raydium.io

Atomic swaps offer a secure, cost-effective, and flexible alternative to traditional centralized exchanges. With atomic swaps, traders can conduct peer-to-peer transactions without relying on intermediaries, maintaining control over their funds while benefiting from lower fees and increased privacy. As the cryptocurrency ecosystem continues to evolve, the adoption of atomic swaps is expected to grow, offering a more decentralized and user-friendly way for traders to exchange digital assets.

Whether you're new to crypto or an experienced trader, atomic swaps should be at the forefront of your trading strategies, enabling you to take full control of your crypto holdings while enjoying the many benefits of decentralized finance.

- Serum https://www.projectserum.com/
- Trader Joe https://traderjoexyz.com/#/

- Ellipsis https://ellipsis.finance/

These are few of the most trusted DEXs that you can choose from that allow you to do atomic swaps for very low network fees. Atomic swaps provide users with the opportunity to maintain anonymous peer to-peer transactions without having to go through a KYC process on a centralized exchange. Users that use atomic swaps are 100% control of their funds unlike centralized exchanges. There is no reason to even really use a centralized exchange if you don't want to. You have the ability to use the many resources that decentralize financing provides such as staking, atomic swaps, lending and borrowing. Are you ready to become your own bank? If so, the atomic swap is your go-to for decentralized freedom.

Chapter 17

Crypto Faucets: The Drip That Keeps Giving

At the time, what may seem valuable today had little monetary worth. Bitcoin was still in its infancy, largely unknown in the early stages of its adoption just like the internet first started, and trading at minimal value.

Gavin Andresen recognized that adoption required experience, not just explanation. By giving small amounts of Bitcoin away for free, he lowered the barrier to entry and allowed people to interact with the technology firsthand. Users could set up wallets, complete simple tasks, and begin understanding how digital currency actually worked.

This marked the birth of crypto faucets — a strategic tool designed to generate awareness and drive early adoption. The model proved effective and was soon replicated across the growing cryptocurrency ecosystem, becoming a foundational method for expanding user participation in the early days of digital assets.

Over time, faucets began to support various cryptocurrencies beyond Bitcoin, such as Ethereum, Litecoin, and other altcoins. The tasks or actions that people needed to complete in exchange for these small amounts of crypto varied, ranging from completing captchas and surveys to watching ads or even playing games. The concept remained the same—provide a way for users to interact with the digital currency ecosystem without requiring an initial investment.

However, it's important to note that while faucets can be a fun and easy way to get involved in crypto, they are unlikely to make you rich. The rewards from most faucets are small, often just a fraction of a cent or a small portion of a cryptocurrency. Faucets work best as a means to introduce new users to crypto or to build a small, growing collection of tokens over time. The key is to accumulate small amounts over time from different sources and let that "drip" effect work its magic, rather than relying on one large source of income.

How Do Crypto Faucets Work?

Crypto faucets are generally simple to use. Most of the time, they require you to complete simple tasks like solving captchas, watching short advertisements, or clicking on a button to claim rewards. After completing the task, you'll be rewarded with small amounts of cryptocurrency. This process is repeated at intervals, sometimes once every few minutes, hours, or even daily, depending on the faucet.

In some cases, users can claim rewards by completing surveys, engaging with games, or participating in referral programs. Faucets often come with a withdrawal threshold, meaning you need to accumulate a minimum amount of cryptocurrency before you can withdraw it to your wallet.

Why Should You Use Faucets?

Here are some reasons why you might want to take advantage of cryptocurrency faucets:

1. **Learn About Cryptocurrencies**: Crypto faucets are an excellent way for beginners to learn about cryptocurrencies and how they work. Faucets provide hands-on experience with receiving and storing crypto, helping users become familiar with wallets, addresses, and transaction processes.

2. **No Investment Required**: Faucets allow you to earn cryptocurrency without having to invest any money. This makes them an attractive option for newcomers who are hesitant to invest in crypto but still want to dip their toes into the ecosystem.

3. **Diversify Your Portfolio**: Faucets support a wide variety of cryptocurrencies, not just Bitcoin. This gives users a chance to accumulate various types of tokens, creating a diversified portfolio over time.

4. **Passive Income**: Faucets can be a form of passive income, allowing users to claim rewards regularly with minimal effort. While it won't replace a full-time job, it can offer a small additional stream of crypto that adds up over time.

5. **Reward for Time Spent**: Faucets reward users for spending time on their platform. The more time you dedicate to interacting with the faucet, the more rewards you can earn. The rewards are typically small, but they can still be a nice bonus for your efforts.

6. **Referral Bonuses**: Many faucets offer referral programs, where users can earn additional cryptocurrency by referring others to the platform. If you have a network of friends or followers who are interested in cryptocurrency, you can potentially earn even more by referring them to the faucet.

Types of Crypto Faucets

1. **Traditional Faucets:** These are the classic faucets where you claim small amounts of cryptocurrency by solving captchas or clicking a button. Examples of traditional faucets include FreeBitcoin, Cointiply, and Moon Bitcoin.

2. **Game Faucets**: Some faucets allow you to earn cryptocurrency by playing simple games. These platforms reward you with small amounts of crypto for completing in-game actions or reaching certain milestones. Examples include Rollercoin and Bitfun.

3. **Survey Faucets**: These faucets allow you to earn cryptocurrency by completing surveys or other tasks. You'll need to answer questions, watch videos, or take part in other activities to

accumulate rewards. Websites like TimeBucks and InboxDollars may offer survey-based crypto rewards.

4. **Micro-Task Faucets**: Micro-task faucets reward users for completing small online tasks such as testing apps, writing reviews, or visiting websites. Some platforms, like FaucetPay, integrate micro-task faucets with their systems, allowing you to earn cryptocurrency for completing various online tasks.

How to Maximize Earnings from Crypto Faucets

1. **Diversify Faucets**: Signing up for multiple faucets increases your chances of accumulating a broader range of cryptocurrencies. It may take some time to manage all the accounts, but the more faucets you're involved in, the more rewards you can earn over time.

2. **Consistency**: Since faucets operate on time intervals, staying consistent and claiming your rewards regularly can help increase your earnings over time. While the payouts may be small at first, consistent claiming adds up.

3. **Referral Programs**: Many faucets offer referral bonuses for getting others to sign up. Sharing your referral links on social media platforms or within crypto communities can boost your faucet income with little extra effort.

4. **Complete All Tasks**: Faucets often offer different types of tasks, and some tasks have higher payouts. Try completing all available activities, from solving captchas to watching product videos or completing surveys, to maximize the amount you earn.

5. **Track Your Earnings**: Use tools like spreadsheets or portfolio trackers to keep track of your faucet earnings. This will help you stay organized and keep an eye on your long-term accumulation.

6. **Avoid Scams**: Since faucets are a target for scams and phishing attacks, always double-check the faucet sites you use. Look for trusted websites with a reputation for paying users and offering reliable payouts. Make sure to read reviews and seek feedback from others in the cryptocurrency community.

Do Your Own Research!

Not all faucets are trustworthy. Some will not give the promised amount, or worse, they simply steal your funds once you've accumulated enough to withdraw. It's crucial to be very careful when choosing which faucets to trust, as the crypto space remains rife with scams and fraudulent websites.

So how can you tell the difference between legitimate crypto faucets and scam websites? Here's a quick guide to help you stay safe while exploring crypto faucets:

1. **Reputation Check**: Always check reviews and feedback from other users. Look for online communities, such as Reddit or Bitcointalk, where real users share their experiences with different faucets. This can help you avoid shady websites.

2. **Transparency**: Legitimate faucets are transparent about their business model. They typically display clear terms and conditions and explain how they generate income to pay users. If the faucet owner is hiding this information or is vague about how they operate, it's a red flag.

3. **Withdrawals**: Before you invest your time into a faucet, look at their withdrawal system. A credible faucet will allow you to withdraw your earnings once a certain threshold is reached. Beware of faucets that have high withdrawal minimums or delay payments.

4. **Task Requirements**: Genuine faucets usually ask users to complete simple tasks, like solving captchas, watching ads, or participating in surveys. If a faucet asks you for sensitive

personal information, like your private keys or wallet recovery phrases, it's likely a scam. Never share your private keys.

5. **Avoid Faucets That Promise Too Much**: If a faucet promises you substantial amounts of crypto for minimal effort, it's probably too good to be true. Crypto faucets that promise huge rewards for little work are often fraudulent and designed to deceive users.

6. **No Payment Without Proof**: If a faucet doesn't show any proof of previous payouts or refuses to share transaction details of successful withdrawals, steer clear. Legitimate faucets will typically provide transaction hashes or other evidence to back their claims.

7. **Referral Schemes**: Many faucets offer referral programs, where you earn a percentage of what your referred users earn. However, if a faucet relies heavily on its referral program and encourages users to promote it rather than focusing on the tasks, it might be trying to build an unsustainable Ponzi scheme.

8. **Time Commitment**: Some faucets will require you to be online for a set amount of time to earn rewards. While this can be a fun way to accumulate small amounts of crypto, if the faucet asks for too much of your time or requests excessive engagement, it may not be worth it.

Best Crypto Faucets to Start Earning

If you're ready to get started with crypto faucets, here are some popular options you can explore:

1. FreeBitcoin (https://freebitco.in): A classic faucet where you can earn Bitcoin by rolling a dice or participating in various activities like referrals, lottery, and more.

2. Cointiply (https://www.cointiply.com): Another well-established faucet that lets you earn Bitcoin by watching ads, completing surveys, and spinning a reward wheel.
3. FaucetPay (https://faucetpay.io): A multi-currency faucet that allows you to earn a variety of cryptocurrencies including Bitcoin, Litecoin, Dogecoin, and more. You can also use FaucetPay as a wallet to manage your crypto earnings.
4. Moon Bitcoin (https://moonbit.co.in): A popular Bitcoin faucet that allows you to claim every 5 minutes. You can accumulate Satoshis over time and even increase your rewards by using a referral link.
5. Bitfun (https://bitfun.co): A faucet that rewards you with Bitcoin for completing simple tasks like viewing ads or participating in games.
6. Rollercoin (https://www.rollercoin.com): A game-based faucet where you can mine Bitcoin by playing mini-games and building your virtual mining farm.
7. FaucetHub (https://faucethub.io): A crypto wallet and faucet aggregation site where you can claim crypto from multiple sources and withdraw to your wallet once you've accumulated enough.
8. TimeBucks (https://www.timebucks.com): A site where you can complete surveys, watch videos, and engage with various activities to earn cryptocurrency, including Bitcoin, Ethereum, and Litecoin.

Crypto faucets may not make you rich overnight, but they provide a simple, fun, and risk-free way to start accumulating cryptocurrency. By participating in multiple faucets and taking advantage of different ways to earn, you can diversify your holdings and build your crypto portfolio over time. Remember, the key to building wealth in the crypto world is to have multiple streams of income, even if those streams are small. Faucets are just one of many ways to get started on your crypto journey, and they offer a low barrier to entry for newcomers who want to explore the world of digital assets. So, why not take advantage of these "drips" and start collecting crypto today?

Good luck and stay safe on your crypto journey!

Chapter 18

Trading Bots: The Future of Trading

Automated trading systems have revolutionized the way financial markets operate. With technology advancing rapidly, the accessibility and capabilities of trading bots have grown immensely. Traders can now leverage artificial intelligence and algorithmic systems to execute complex trades at lightning speeds, enabling them to take advantage of market opportunities that would be impossible for human traders to capture in real-time.

The concept of algorithmic trading continues to evolve, with modern trading bots now capable of analyzing vast amounts of data, adapting to changing market conditions, and executing trades without human intervention. These bots can monitor multiple markets simultaneously, making decisions based on technical indicators, historical data, and even news sentiment. This has significantly reduced the time between decision-making and execution, allowing for more efficient and profitable trading strategies.

Automated trading systems and bots are transforming the world of trading by making it more efficient, accessible, and profitable. By understanding the history and development of this technology, as well as its benefits and potential risks, traders can make informed decisions and take full advantage of the opportunities that automated trading offers. Whether you're a novice looking to enter the world of trading or an experienced trader seeking to improve your strategies, automated trading could be the tool you need to take your trading game to the next level.

In recent years, automated trading systems have gained significant popularity, revolutionizing how investors and traders approach markets. These systems have advanced rapidly with improvements in IT and AI technologies, and are now integral parts of global financial markets, including crypto exchanges.

Back in 2014, it was reported that more than 75% of all stock shares traded on U.S. exchanges came from automated trading systems. Over time, these systems have not only dominated traditional markets like stocks and forex, but they have also expanded into cryptocurrency markets, making crypto trading more accessible and efficient.

How do Crypto Trading Bots Work?

Crypto trading bots function as programs designed to automate the process of trading cryptocurrencies on behalf of investors. Normally, manual trading requires investors to constantly monitor the markets, analyze trends, and make decisions on which cryptocurrency to buy or sell. With crypto trading bots, however, this process is fully automated. The bots collect market data, analyze trends, assess potential risks, and then execute trades based on predefined parameters.

The bots work using algorithms that are tailored to specific market conditions, eliminating human errors such as missed opportunities, poor risk management, or emotional decisions. Whether it's a simple technical indicator, like a moving average crossover, or a complex strategy involving multiple factors, crypto trading bots can manage trades efficiently around the clock, without requiring the trader to be physically present in front of a computer screen.

The four key components that enable crypto trading bots to function efficiently are:

1. **Signal Generator**: This is the primary function of the bot. It analyzes market data, predicts trends, and identifies potential trades. The signal generator relies on technical analysis indicators to make trading decisions, ensuring that the trades are made based on data rather than emotional impulses.

2. **Risk Allocation**: Risk management is critical in trading, and the bot ensures that capital is allocated in a way that reduces risk according to predefined rules set by the trader. For example,

   it can adjust the percentage of capital to invest in each trade or apply stop-loss measures automatically to minimize potential losses.

3. **Execution**: Once a trade is identified by the signal generator, the execution component takes over. It sends the trade orders to the cryptocurrency exchange via API keys, instructing the system to buy or sell assets according to the parameters set by the user. The execution is typically fast and precise, allowing the user to take advantage of opportunities that may otherwise be missed.

4. **Arbitrage Bots**: These bots are specialized in spotting price discrepancies across multiple exchanges. They allow traders to exploit price differences for the same asset between exchanges and make a profit. By executing trades across different exchanges at the right time, arbitrage bots can capture quick profits from the price variation.

With crypto trading bots, users no longer have to spend hours monitoring the markets. These bots run 24/7, executing trades as needed and allowing investors to trade globally, regardless of time zone or location. They also remove much of the emotional element from trading. Since trading decisions are made based on pre-configured rules, the bot eliminates the influence of fear or greed that can negatively impact human traders.

Crypto trading bots have become an essential tool for both beginner and experienced traders looking.

to automate their trading strategies and take advantage of the market efficiently. By automating trades, these bots provide the opportunity for traders to make consistent profits without constantly being in front of their computers.

Exploring Trading Bot Strategies: Diversifying Your Portfolio

In the world of cryptocurrency trading, automated trading bots have become an essential tool for traders looking to optimize their strategies and profits. Below are three popular trading bot strategies you can implement individually or combine them to diversify your portfolio:

1. **Arbitrage Bots**: An arbitrage bot is programmed to identify price discrepancies across different exchanges and execute trades to profit from those gaps. For example, if Bitcoin (BTC) is trading for $10,000 more on Binance compared to KuCoin, the bot can automatically purchase BTC on the lower-priced exchange (KuCoin) and sell it on Binance for a quick profit. Arbitrage bots leverage market inefficiencies and execute trades faster than humans, making them highly effective in capturing price gaps.

2. **Grid Bots**: Grid bots are simple yet effective, making them ideal for beginners. The strategy involves setting multiple price levels, or "grids," above and below the current market price of a cryptocurrency. The bot will then automatically buy and sell at predetermined price intervals, accumulating small profits from each trade. This method benefits from market fluctuations, capitalizing on price movements within a range.

3. **Sandwich Bots**: Sandwich trading bots detect when a trader places a bid on a cryptocurrency and based on this information, place a higher bid on the same asset. The bot essentially "sandwiches" the trader's order, betting that they will still want the asset. Once the bot wins the cryptocurrency, it immediately resells it to the original bidder at a higher price, exploiting the difference for a profit.

This strategy takes advantage of market momentum and user behavior.

Key Benefits of Using a Trading Bot

Utilizing trading bots offers numerous advantages, especially when it comes to efficiency and execution speed. Below are five key benefits that make trading bots a valuable asset to any trader:

1. **Emotion-Free Trading**: Emotions such as fear and greed can be detrimental to a trader's success. Research shows that over 80% of private traders lose money due to emotional decisions that cloud their judgment. The cryptocurrency market is volatile, and emotional trading can lead to significant losses. By using a trading bot, you eliminate emotional interference from the decision-making process. Bots follow pre-programmed strategies, ensuring a systematic and disciplined approach. Instead of succumbing to emotions, you can be part of the 20% of traders who profit from trading bots.

2. **Speed and Efficiency**: In the trading world, speed equals profit. Trading bots operate much faster than humans, executing trades in nanoseconds. While it may take a human several minutes to manually complete a trade, a bot can perform thousands of calculations and transactions in the same amount of time. This immediate response to changing market conditions ensures you never miss an opportunity. With bots, you can enter and exit trades more quickly, potentially increasing your profits or minimizing losses.

3. **Backtesting and Paper Trading**: Just as pilots use flight simulators to practice and refine their skills, traders can use market simulators to test their strategies without risking actual capital. Backtesting allows you to evaluate the performance of a trading strategy using historical data, helping you understand how the bot would have reacted in past market conditions. Paper trading, which involves executing trades in a simulated environment, enables traders to practice risk-free.

These tools are essential for refining strategies before deploying them in live markets, and the best part is that many backtesting and paper trading services are free to use.

4. **Round-the-Clock Trading**: One of the most significant advantages of trading bots is their ability to trade 24/7. The cryptocurrency market never sleeps, and human traders often find it challenging to monitor the market constantly. Bots, however, can operate around the clock, seizing opportunities at any time of day or night. This constant vigilance allows for more frequent and diverse trade, increasing the chances of capitalizing on profitable market movements.

5. **Customization and Flexibility**: Trading bots can be tailored to suit your individual preferences and strategies. Whether you want to set complex conditions for entry and exit points or simply want to automate a basic strategy, bots offer unparalleled flexibility. You can choose from pre-built strategies or create your own, and as your trading experience grows, you can adjust the bot's settings to fit your evolving needs. This level of customization ensures that you're always in control of your trading approach.

Consistency and Discipline: Key to Success in Trading

You've likely heard the saying "consistency is key" — and in trading, it's especially true. With a trading bot, once you set the rules, execution happens automatically, ensuring that the bot adheres to your plan even in volatile markets. Emotions like fear of loss or the temptation for extra profits often lead to inconsistent decisions, disrupting a trader's strategy. Automated bots eliminate this risk by executing trades strictly according to the pre-programmed strategy, free from human emotional interference.

Moreover, human errors are minimized. For example, a bot won't mistakenly place a buy order for 500 shares as a sell order for 5,000 shares. One of the greatest challenges traders face is sticking to their plan, and trading bots ensure that discipline is maintained. It's essential to follow the plan, even when losses occur—since losses are an inevitable part of trading. When traders deviate from their rules or become emotionally driven after a few losing trades, they often miss opportunities for future success. Automated systems help traders stay disciplined and consistent, ultimately improving their long-term results.

Diversifying Your Trading Portfolio with Bots

The old saying, "Don't put all your eggs in one basket," couldn't be more relevant in trading. With automated trading systems, you can manage multiple accounts or use various strategies simultaneously. This enables you to diversify your portfolio, spreading risk across different instruments and exchanges while simultaneously hedging against potential losses. What would be incredibly difficult or time-consuming for a human to manage, such as monitoring multiple markets and executing trades, is effortlessly handled by a bot in milliseconds.

The ability to scan opportunities across different markets, generate orders, and track ongoing trades means that automated trading systems can enhance your portfolio's efficiency and effectiveness.

The Drawbacks of Trading Bots

Despite their many advantages, trading bots also have potential drawbacks that should be considered:

1. **Mechanical Failures:** Technology, though powerful, is not infallible. Trading bots might seem simple at first glance—set the software, program the rules, and let it run. However, in reality, they are sophisticated systems that can fail. For instance, if an internet connection is lost, the bot may fail to execute trades, or it might not send orders to the market at all. A bot

might also be affected by system crashes or glitches in the platform. It's crucial to have backup plans and ensure that everything works smoothly.

2. **Over-Reliance on Automation**: Trading bots are designed to execute predefined rules, which means that they operate on the principle of automation without human intervention. While this can be beneficial, it can also lead to a lack of flexibility. Market conditions can change rapidly,

   and a bot might not always be able to adapt to unforeseen situations in the way a human trader might. Over-reliance on a bot without understanding the fundamentals of trading could lead to poor decision-making.

3. **Costs and Fees**: While many trading bots come with advanced features and automation, they often come with subscription or usage fees. Some services charge based on the number of trades executed, while others might take a percentage of profits. These costs can accumulate over time, reducing overall profitability. As a result, it's important to weigh the potential costs against the benefits before committing to an automated trading service.

Maximizing Your Potential with Cryptocurrency Trading Bots

Trading bots are powerful tools designed to help you minimize human error and increase speed in executing trades—two critical factors for success in the fast-paced world of cryptocurrency trading. When set up correctly and executed with discipline and a clear game plan, trading bots can significantly enhance your potential in the market. There are numerous trading bots available, some offering affiliate programs for referring others, but it's essential to understand that the most reliable options usually come with a paid subscription. Free bots often lack the advanced features and support necessary for consistent success. As with many things in life, if you want a system that works, you'll need to invest in it.

Here are some popular cryptocurrency trading bots to consider at the time of writing:

- Pionex
- Kucoin
- 3Commas
- CoinRule
- Shrimpy
- CryptoHopper
- Trality
- Bitsgap
- Quadency
- CryptoHero
- TradeSanta
- HAASONLINE
- Stoic

With the advent of AI and blockchain technology, cryptocurrency traders can now leverage the same tools that institutional investors have had access to since the 90s. Trading bots eliminate human errors, and the speed at which they can execute trades—sometimes in nanoseconds—vastly outperforms manual trading.

By diversifying assets and using bots effectively, you can potentially earn significant profits.

From my personal experience, I have made hundreds of thousands of dollars using trading bots, and I've helped friends achieve similar success with strategies like leverage and arbitrage trading. For cryptocurrency traders, the arbitrage trading bot is an especially valuable tool for maximizing returns by capitalizing on price differences between exchanges.

The cryptocurrency market is already moving fast, and most active traders are using bots. Why not level the playing field by setting up your own automated trading system? Trading bots remove the emotional component from decision-making, which is a common pitfall for manual traders. With AI technology becoming more sophisticated, trading bots are able to analyze the market and execute trades faster and more effectively than humans can, often with better accuracy. In fact, some bots boast a remarkable win rate of 75-86%, making them an attractive choice for serious traders.

It's essential to do your own research (DYOR) and select a trading bot that aligns with your unique trading strategy. Not all bots are created equal, and they may perform differently depending on the exchange you use. Take the time to find a bot that integrates seamlessly with your preferred platform and begin developing a structured plan to maximize your chances of success. By doing so, you can position yourself to become a professional trader, capitalizing on the immense opportunities the crypto market offers.

Chapter 19

Dapps, Metamask, and Walletconnect: Connecting to the Blockchain

As you venture into the cryptocurrency markets, it's essential to equip yourself with the right tools to ensure success in buying, selling, or even trading cryptocurrencies. Additionally, these tools will enable you to engage in decentralized finance (DeFi) activities like purchasing NFTs or participating in play-to-earn games. In this chapter, we aim to educate you on the necessary tools to connect with decentralized applications (DApps) on various crypto exchanges.

These tools will allow you to do everything from buying cryptocurrencies to staking and more. However, it's important to note that these tools are primarily for use with decentralized exchanges, not centralized ones. Centralized exchanges are controlled, meaning they hold your private keys and thus don't offer you complete control over your funds.

When setting up your wallet on platforms like MetaMask or any other wallet that provides you with a 12-word recovery phrase, it's crucial to write this phrase down. Never save it on your computer or phone, as doing so exposes you to the risk of losing your funds if someone gains access to it. To protect your private keys, write down your 12-word recovery phrase multiple times and store it securely in a safe place— this will ensure your crypto holdings remain secure.

Now, let's discuss what decentralized applications (DApps) are. But first, it's essential to understand their history, as tools like MetaMask and WalletConnect were created in response to the rise of DApps. There are many other wallets that work with DApps, such as the Binance Chain Wallet, Icon, and Phantom These wallets enable you to connect to various decentralized exchanges (DEXs) and perform transactions. However, just like MetaMask, these wallets require you to set up a wallet and record your 12-word recovery phrase before you can access them and connect to the exchanges.

A Brief History of Decentralized Applications (DApps)

Decentralized applications (DApps) are applications built on decentralized networks. Unlike traditional applications, which are owned and operated by a single corporation or individual, DApps run on peer-to-peer networks using blockchain technology. This structure provides users with more control over their data and ensures greater privacy for their transactions. Additionally, DApps eliminate the need for centralized intermediaries, making them more secure and private.

Each DApp is powered by smart contracts that enable them to function on a specific blockchain. This decentralized design ensures that the DApp cannot be taken down or altered by human errors. Once a smart contract is deployed, users can interact with the DApp, even if part of the network is down.

Bitcoin, which launched in 2009, was the first cryptocurrency to establish the foundation for a new era of peer-to-peer financial services, marking a break from traditional centralized services. However, it wasn't until 2014 that David Johnston's paper, *The General Theory of Decentralized Applications (DApps)*, laid out a formal definition for DApps and introduced the concept of decentralized applications to the world.

In 2015, Ethereum revolutionized the blockchain landscape by launching the first smart contract-powered blockchain. This allowed projects to be built on Ethereum's network, facilitating decentralized applications and more complex functionality, particularly for finance. Ethereum's introduction made it possible to execute decentralized applications that went beyond simple transactions.

One of the first DApps to gain widespread attention was *CryptoKitties*, a game that launched on Ethereum in 2017. The game's popularity briefly overwhelmed Ethereum's network, with users experiencing delays of several days due to a backlog of over 30,000 pending transactions. This highlighted the limitations of Ethereum's network at the time, but it also marked a significant moment in the history of decentralized applications.

Since 2017, the world of decentralized applications has expanded exponentially, with new DApps emerging across multiple industries. The evolution of blockchain technology continues to drive the growth of decentralized finance, gaming, NFTs, and more, giving users greater control and access to innovative services.

Decentralized applications (Dapps) have revolutionized how we interact with digital, platforms especially in the world of finance, gaming, and trading. While these apps are built on decentralized networks, they solve numerous problems that traditional applications face, which we'll discuss here. Let's explore the three main issues that traditional applications commonly face, and how Dapps are solving them:

1. **Centralized Control & Data Privacy Issues:** In traditional centralized applications, a single entity (a company or service provider) controls all data and user interactions. This creates concerns regarding data privacy and security. Since users don't have control over their data, it becomes vulnerable to hacks, breaches, and misuse. Companies that control this data also have the authority to sell it to third parties or use it for their gain, often without user consent.

   How Dapps Solve It: Dapps eliminate the need for a central authority, allowing users to have full control over their data. Since they operate on decentralized networks (blockchains), they provide transparency, privacy, and security to users. Smart contracts ensure that no one can alter or delete the data, and all transactions are publicly visible for verification. This decentralized nature of Dapps ensures that users own their data and can interact with applications without the fear of their information being exploited.

2. **Single Points of Failure:** Centralized applications rely on centralized servers. This means that if the servers go down, the whole application becomes inaccessible to users, resulting in downtime and service interruptions. Traditional apps are prone to technical failures, which can disrupt services, prevent users from completing transactions, or cause a loss of data.

How Dapps Solve It: Dapps are built on blockchain networks, which are distributed and decentralized across numerous nodes. This decentralized architecture means that there's no single point of failure. Even if part of the network goes down, the rest of the blockchain keeps running smoothly, ensuring uninterrupted access and functionality. The blockchain also makes sure that all information is replicated across the network, preventing data loss due to server failure.

3. **Trust and Intermediary Dependency:** Traditional applications, especially those in finance, rely heavily on intermediaries (banks, brokers, payment processors, etc.) to facilitate transactions. These intermediaries introduce fees, delays, and reliance on trust. In situations like cross-border payments or online purchases, intermediaries can cause delays, add extra costs, and sometimes even be a point of failure in the transaction process.

How Dapps Solve It

Dapps use smart contracts to execute and enforce transactions directly between users. Since the blockchain is a trustless network, no intermediary is required for validation. Smart contracts automatically execute the agreed-upon terms when certain conditions are met, removing human errors, delays, and middlemen from the process. This results in faster, cheaper, and more secure transactions, enabling peer-to peer (P2P) services without the need for traditional financial intermediaries.

Decentralized applications (Dapps) address significant problems that traditional applications face, including centralized control, single points of failure, and the reliance on intermediaries. By harnessing blockchain technology and smart contracts, Dapps empower users to control their data, ensure transparency, and engage in secure, direct transactions without intermediaries. As Dapps continue to evolve, their widespread adoption has the potential to redefine how we interact with the digital world, offering benefits that traditional applications simply can't match.

In the next chapters, we'll dive deeper into some of the best Dapp platforms and how you can utilize them to maximize your potential in the cryptocurrency and NFT space.

Problems with Traditional Apps

1. **Data Misuse:** One of the primary concerns with traditional apps is the potential for data misuse. Many apps collect far more personal information than users may be aware of or comfortable with, often without obtaining proper consent. This information can then be sold to third-party vendors, or, in the worst-case scenario, it could be exposed in the event of a data breach. The lack of control over how your data is used is a significant risk in today's digital landscape, leaving users vulnerable to exploitation and theft.
2. **Vendor Lock-In:** Another common issue with traditional applications is the concept of vendor lock-in. This occurs when users become "locked" into a particular app or service due to longterm contracts or because the app makes it exceedingly difficult to switch to another provider. For example, users may find themselves stuck with a subpar service because canceling their subscription is either too expensive or complicated. Additionally, these apps often restrict your ability to transfer your data or settings to another platform, leaving you with little flexibility and no control over your own digital footprint.

3. **Censorship:** In some countries, such as China or Russia, censorship can severely limit what users are allowed to access and share on traditional platforms. Governments and corporations may restrict or control what content is allowed on social media, preventing free speech and blocking access to certain services. Social media platforms themselves may be pressured to comply with local regulations, resulting in the removal or restriction of content that doesn't align with governmental views. In extreme cases, entire platforms may be banned, leaving users without access to key communication tools. This suppression of free speech and censorship of information highlights the fragility of individual freedom in centralized systems.

The Power of Decentralized Applications: Breaking Free from Traditional Control

The rise of Decentralized Applications (Dapps) has the potential to disrupt traditional models of control and regulation that have been dictated by private corporations and governments. These applications are designed to function without centralized authority, empowering users and removing many of the barriers imposed by traditional systems. Below are six key benefits of Dapps that address issues prevalent in conventional applications:

1. **Open Source:** One of the defining features of many Dapps is that their code is open source, often made publicly available on platforms like GitHub. Unlike traditional apps, where the internal workings are hidden, Dapps allow anyone with the necessary programming skills—or the willingness to learn—to inspect the code. This transparency ensures that there are no hidden bugs, backdoors, or malicious features lurking within the application. Open-source code fosters an environment of continuous improvement and innovation, where developers can collaborate and contribute to the growth of the ecosystem. Additionally, if a Dapp doesn't meet expectations, the community can easily copy the code and launch a better version, known as a hard fork.

2. **Trustless Transactions:** In traditional business models, contracts or written agreements are necessary to ensure that all parties adhere to their commitments. The foundation of these contracts is trust—trust that the other party will uphold their end of the deal. With Dapps, however, trust becomes an unnecessary factor. These applications utilize smart contracts, which are self-executing contracts where the terms are directly written into code. Once a condition is met, the contract is automatically executed, and there is no need for a third party to enforce it. A prime example is decentralized exchanges, where you can swap cryptocurrencies without needing to trust the exchange itself. The trust lies in the blockchain and the code that runs it, providing a level of security and automation that traditional systems can't match.

3. **Enhanced Security:** By operating on decentralized networks, Dapps are inherently more secure than centralized applications. Since the data is distributed across multiple nodes on the blockchain, it becomes nearly impossible for a single point of failure to compromise the entire system. Unlike traditional apps, which store user data on centralized servers that can be targeted by hackers, Dapps leverage the blockchain's encryption and consensus mechanisms to protect users' information. This distributed nature of Dapps significantly reduces the risk of data breaches, hacking, and identity theft.

4. **No Intermediaries:** One of the primary issues with traditional applications is the reliance on intermediaries. Whether it's a bank, a payment processor, or even a social media platform, intermediaries act as gatekeepers and often take a cut of your earnings or control how you interact with services. Dapps remove the need for these intermediaries by allowing peer-to-peer interactions, where users can directly transact with one another without a middleman. This can be especially beneficial in the world of finance, where decentralized finance (DeFi) applications

allow users to access financial services such as lending, borrowing, and earning interest without relying on banks or financial institutions.

5. **Censorship Resistance:** Censorship is a growing concern in many parts of the world, especially with regard to social media platforms and content-sharing apps. Governments or private corporations may impose restrictions on what users can access or share. Dapps, by their very nature, are resistant to censorship. Since they run on decentralized networks, there is no central authority that can control or block access to the service. This ensures that users in censorship-heavy regions can still participate freely in activities like social networking, financial transactions, and online gaming without fear of restrictions.

6. **Privacy and User Control:** Traditional apps often require users to provide extensive personal information, which can be sold to third parties or misused without their consent. Dapps, on the other hand, allow users to retain full control over their data. With decentralized applications, users can choose what information to share and how to interact with the platform. Privacy is baked into the system, as blockchain technology ensures that data is encrypted and stored in a way that prevents unauthorized access. In a decentralized environment, users are not subjected to the whims of thirdparty data brokers or companies looking to exploit their personal details for profit.

Key Features of MetaMask

MetaMask users can seamlessly connect to a variety of decentralized finance (DeFi) platforms, token exchanges, and other Ethereum-based applications. Here are some key features:

1. **Private Key Management**: MetaMask provides users complete control over their private keys, which are stored locally on their device. The wallet does not have access to or store your keys, ensuring that you are the only one who can access and control your funds.

2. **Browser Integration**: As a browser extension, MetaMask works with browsers like Google Chrome, Firefox, and Brave. This integration makes it easy to interact with dApps directly, without needing to leave your browser window. You can access everything from decentralized exchanges (DEXs) to NFT marketplaces and DeFi lending platforms.
3. **Token Management**: MetaMask supports all ERC-20 tokens, as well as ERC-721 tokens (NFTs), allowing users to send, receive, and manage a wide variety of Ethereum-based assets. You can view your token balance and transaction history directly within the MetaMask interface.
4. **Cross-Platform Functionality**: MetaMask now supports both desktop and mobile devices. The app works across browsers (Chrome, Firefox, Brave) and also has mobile apps for iOS and Android, making it accessible and easy to use from anywhere.
5. **User-Friendly Interface**: MetaMask is designed to be simple and intuitive, making it easy for both beginners and experienced users to interact with Ethereum-based applications. Its design focuses on a streamlined user experience while providing robust functionality for advanced users.
6. **Security**: MetaMask allows users to encrypt their wallet with a password. It also supports seed phrase recovery, which ensures that you can always recover access to your wallet even if you lose your device.

How MetaMask Works with DApps

When you connect MetaMask to a decentralized application, the application requests access to your wallet's Ethereum address. This allows you to sign transactions and interact with smart contracts directly from the dApp interface, while MetaMask handles the underlying transaction processes. MetaMask serves as a bridge between the user and the decentralized world, simplifying interactions with the blockchain by abstracting technical complexities.

Connecting MetaMask to DApps:

- Upon visiting a dApp's website (for example, a decentralized exchange or NFT marketplace), you'll usually be prompted to connect your MetaMask wallet.
- Once connected, you can perform actions such as making swaps on decentralized exchanges, minting or purchasing NFTs, and staking tokens on DeFi platforms—all without needing to trust third-party intermediaries.

Security Considerations

While MetaMask provides strong security features, users must remain vigilant. For instance,

phishing attacks and fake MetaMask apps can lead to the theft of your private keys. Always make sure you're using the official MetaMask extension or app, and never share your seed phrase with anyone.

WalletConnect and Binance Smart Chain (BSC)

Now that we've covered MetaMask, let's talk about WalletConnect and Binance Smart Chain (BSC). WalletConnect is an open-source protocol that enables users to connect their wallets to decentralized applications (dApps) through a secure QR code scanning process. Unlike MetaMask, WalletConnect is not tied to a specific wallet but rather serves as a universal bridge for a variety of wallets. By scanning a QR code, WalletConnect establishes a secure connection between your wallet and the dApp you're trying to use.

Some of the most popular wallets that integrate with WalletConnect include Trust Wallet, Argent, and more.

Binance Smart Chain (BSC) is a blockchain network developed by Binance to support the creation

and execution of decentralized applications (dApps). It features faster transaction speeds and lower fees than Ethereum, making it an attractive alternative for developers and users looking to access dApp services more cost-effectively. BSC also supports decentralized finance (DeFi) protocols, allowing users to swap tokens, yield farm, and participate in liquidity mining directly through the dApp ecosystem on BSC.

In conclusion, MetaMask, WalletConnect, and Binance Smart Chain are powerful tools that provide users with the ability to interact with decentralized applications and the broader blockchain ecosystem. Each one offers unique features, but when combined, they give you the flexibility to access a wide variety of dApps and manage your assets across different blockchains. Whether you're managing Ethereum tokens, engaging in DeFi, or exploring NFT marketplaces, these decentralized applications (dApps) are reshaping how we interact with the world of cryptocurrency.

Binance Smart Chain (BSC) and Binance Chain Wallet: A Comprehensive Overview

The Binance Smart Chain (BSC) is a blockchain network created by Binance, one of the largest cryptocurrency exchanges in the world. It was launched in 2019 with a focus on providing fast, low-cost, and scalable solutions for decentralized applications (DApps) and decentralized finance (DeFi). BSC was designed to overcome the limitations of Ethereum's blockchain, particularly in terms of transaction speed and fees, making it a popular choice for developers and users in the blockchain ecosystem.

Key Functions of Binance Smart Chain

1. **Sending and Receiving Cryptocurrencies:** Binance Smart Chain facilitates the secure and fast transfer of cryptocurrencies, including its native currency, BNB (Binance Coin). Users can send and receive assets on BSC with minimal transaction fees, ensuring a cost-effective method for transferring digital assets.

2. **Token Issuance and Supply Management:** BSC allows users to control token issuance and supply by using functions such as minting, burning, locking, unlocking, freezing, and unfreezing funds. This functionality provides a high level of flexibility in managing tokens on the Binance Smart Chain network.
3. **Issuance of New Cryptocurrencies:** New cryptocurrencies can be issued on Binance Smart Chain based on the BEP-2 token standard. An example of this is BNB, which exists as a BEP-2 token on the Binance Smart Chain network. BSC's ability to issue tokens makes it an ideal environment for launching various types of digital assets, including stablecoins, governance tokens, and more.

The Binance Chain Wallet

The Binance Chain Wallet is the official wallet for managing Binance-based cryptocurrencies. This wallet provides access to multiple blockchains, including Binance Chain, Binance Smart Chain, and Ethereum, making it a versatile tool for users engaged in the cryptocurrency ecosystem. It allows users to securely manage their assets, store private keys, and interact with decentralized applications (DApps) on various blockchains.

Key Features of the Binance Chain Wallet

1. **Private Key Management:** The Binance Chain Wallet gives users full control over their private keys, ensuring they are the sole custodians of their digital assets. By maintaining

control over private keys, users can secure their funds and avoid relying on centralized exchanges, where funds may be at risk of hacking or theft.

2. **Secure and Decentralized Transactions:** This wallet enables users to send and receive funds across various blockchain networks, including BSC and Ethereum. With its decentralized nature, Binance Chain Wallet offers quick, secure, and private transactions without the need for intermediaries. It ensures users can trade and swap digital assets in a way that respects their privacy and autonomy.

3. **Low-Cost and Fast Transactions:** Binance Smart Chain operates using the Tendermint Byzantine Fault Tolerance (BFT) consensus mechanism, which ensures fast and efficient transaction processing. BSC is designed to handle up to 300 transactions per second (TPS), significantly outperforming Ethereum in terms of transaction speed. Additionally, transaction fees on Binance Smart Chain are usually just a few pennies, making it much more cost-effective than Ethereum, where gas fees can be prohibitively high.

4. **Smart Contract Compatibility:** Binance Smart Chain is fully compatible with Ethereum-based

   DApps, allowing developers to port their existing applications to BSC with minimal changes. This compatibility with Ethereum's EVM (Ethereum Virtual Machine) enables users to access a wide range of decentralized applications, including decentralized exchanges (DEXs), lending protocols, yield farming platforms, and more.

5. **Cross-Blockchain Support:** The Binance Chain Wallet supports a variety of blockchain networks, allowing users to manage their assets across Binance Chain, Binance Smart Chain, and Ethereum. This multi-chain support enables users to participate in different ecosystems and interact with a broad range of DApps and services.

6. **User-Friendly Interface:** Available as a browser extension for popular browsers like Chrome, Brave, and Firefox, as well as a mobile app for iOS and Android, the Binance Chain Wallet provides a seamless and user-friendly experience. Its design makes it easy for users to manage their crypto holdings, access decentralized exchanges, and execute smart contract functions with minimal hassle.

Binance Smart Chain's Advantages Over Ethereum

1. **Transaction Speed and Cost Efficiency:** One of the primary benefits of Binance Smart Chain is its ability to process transactions at a much faster rate than Ethereum. While Ethereum's proofof-work (PoW) system often leads to long transaction confirmation times and high gas fees, BSC's Proof of Stake (PoS) model ensures rapid transaction processing and minimal fees. This makes BSC an ideal blockchain for applications that require high throughput and low-cost transactions, such as DeFi platforms, NFT marketplaces, and DEXs.

2. **Scalability:** Binance Smart Chain is designed to scale efficiently, allowing it to handle a large volume of transactions without experiencing congestion. Its block time of just 3 seconds and its ability to process up to 300 TPS make it suitable for applications with high user activity, ensuring that the network remains responsive even during periods of high demand.

3. **Ethereum Compatibility:** While BSC offers faster and cheaper transactions, it also maintains compatibility with Ethereum-based applications. This enables developers to leverage Ethereum's existing ecosystem while benefiting from BSC's superior performance. The compatibility ensures that users can interact with Ethereum-based DApps on Binance Smart Chain, broadening the opportunities for users and developers alike.

4. **Ecosystem Growth:** The Binance Smart Chain ecosystem has grown rapidly since its launch, with numerous DeFi projects, DEXs, and NFT platforms being built on the network. As more

projects migrate to BSC due to its low fees and high performance, the ecosystem continues to expand, providing more opportunities for users to participate in decentralized finance, yield farming, and other blockchain-based services.

Binance Smart Chain has established itself as a strong competitor to Ethereum by providing faster transaction speeds, lower fees, and enhanced scalability for decentralized applications. With its Proof of Stake (PoS) consensus mechanism and Ethereum compatibility, BSC offers developers and users a more efficient and cost-effective alternative to Ethereum. The Binance Chain Wallet plays a crucial role in providing users with easy access to BSC and other blockchains, allowing them to manage their assets, participate in DeFi activities, and interact with a wide range of DApps securely and privately. As the blockchain ecosystem continues to evolve, Binance Smart Chain's low-cost, high-speed infrastructure will likely continue to attract developers and users looking for an optimized experience in decentralized finance and beyond.

Transactions on Binance Smart Chain (BSC) are processed in seconds and cost less than $0.04 per transaction, offering a significant advantage over Ethereum, where fees can surge from $100 to $300 per transaction during periods of network congestion. This disparity is due to Ethereum's reliance on Proof of Work (PoW) consensus, which limits transaction speeds. In contrast, BSC leverages a faster and more scalable architecture, processing around 300 transactions per second. As a result, Binance Smart Chain is a preferred choice for decentralized finance (DeFi) applications, staking, high-yield farming, and exchanges. Notably, popular wallets like MetaMask, Trust Wallet, and hardware wallets like Ellipal are compatible with the Binance Smart Chain ecosystem.

Now that we've explored some of the top decentralized applications (Dapps) such as MetaMask for Ethereum-based tokens, Binance Chain Wallet for BNB-based tokens, and WalletConnect for connecting to a wide variety of Dapps, let's take a look at other notable Dapps across different blockchains. If you're interested in exploring further, the State of the Dapps website provides an extensive directory of decentralized applications.

Popular Dapps by Blockchain

Ethereum:

- OpenSea – A leading peer-to-peer marketplace for NFTs and digital assets.
- Oasis – A decentralized finance (DeFi) platform for borrowing DAI against collateral.
- Chainlink – An oracle that bridges the gap between smart contracts and real-world data.
- Decentraland – A virtual reality platform where users can purchase virtual real estate and interact with others in a 3D space.
- Uniswap – An automated market maker (AMM) protocol enabling token swaps without the need for a central exchange.

EOS:

- Upland – A virtual property game that allows players to purchase, trade, and collect virtual properties tied to real-world addresses.
- Yup – A social platform that rewards users for rating and sharing opinions on various items or experiences.
- Prospectors – A multiplayer strategy game where players mine gold and earn crypto within a simulated 19th-century environment.
- Blockchain Cuties – A collectible crypto game featuring unique, tradable characters in the form of NFTs.

ICON:

- ICON Governance – A platform enabling decentralized governance and decision-making on the

  ICON blockchain.
- GangstaBet – A digital collectible platform where users can engage in various decentralized betting games.
- OMM – A decentralized money market for borrowing, lending, and earning interest on various

  assets.

Additional Blockchain Ecosystems and Dapps

- Solana: Known for its high-speed transactions and low fees, Solana hosts a growing array of DeFi projects, NFT platforms, and more.
- Polkadot: A blockchain designed to connect various blockchains, enabling interoperability and a range of innovative Dapps across multiple chains.
- Avalanche: A platform that offers high-throughput and low-latency consensus, supporting decentralized finance, gaming, and asset management Dapps.

Whether you're exploring NFTs, DeFi applications, gaming platforms, or governance tools, the Dapp ecosystem continues to expand, offering countless opportunities for users to engage with blockchain technology in various ways.

Some of the most popular uses for Dapps currently use today are mentioned below:

- **Trading**: Dapps like Uniswap, Sushiswap, SpookySwap, PancakeSwap and Aave allow users to trade their cryptocurrencies safely and anonymously.

- **Social Media**: Social media has also gotten the Dapps interaction! Including Steemit, Kin and Hive a popular social blogging platform.
- **Gaming**: There is a huge range of Dapps based games, including Decentraland, Sandbox, AnRKey X, Blockchain Cuties, Townstar and My Neighbor Alice.
- **Entertainment**: The Dapps industry also has its own YouTube equivalent, known as DTube, Minds, and Livepeer; a blockchain-powered video sharing platform.

Unlike traditional apps, which usually need to be downloaded, purchased and installed before they can be used (unless they are web apps), Dapps are generally accessed through Dapps browser (like MetaMask or Brave Browser or even Google Chrome) and don't need to be installed.

Decentralized applications (DApps) are rapidly becoming more accessible and increasingly beneficial, especially for individuals in countries with stringent regulations or censorship. In these regions, where access to social media platforms and websites is restricted, DApps offer an open-source alternative. Users now have the freedom to express themselves through blogs and gain access to apps that might otherwise be banned or restricted in their home countries.

The power of DApps lies in their ability to provide anonymous, peer-to-peer transactions and interactions with the world without requiring approval from governments or private corporations. They empower users to freely watch, purchase, and communicate without fearing the oversight of controlling entities. With DApps, individuals can exercise their freedom of speech and engage in financial activities without concern for censorship or surveillance. This is the true strength of decentralized finance (DeFi).

For those reading this chapter, I encourage you to start moving away from centralized banking and social media platforms. Today, there are applications that allow you to speak freely on any topic, without censorship. The best part is, you can even earn cryptocurrency by using these platforms! Isn't that amazing?

By embracing blockchain technology and smart contracts, individuals who take the time to learn about them can create financial independence while connecting to an ecosystem designed to support growth and provide a space where your opinions are heard—without judgment.

So, what are you waiting for? It's time to cancel those subscriptions, end censorship, and explore the exciting world of decentralized applications. I look forward to meeting you there!

Chapter 20

Hardware Wallets The Power of Cold Storage

What if there was a way to keep your funds completely offline, away from hackers, while maintaining complete anonymity without compromising security? In this chapter, we will delve into the history of hardware wallets, explore their benefits, and discuss some of the top hardware wallets available on the market today. This chapter is particularly valuable for those currently relying on online wallets (also known as hot wallets) and looking to transition to cold storage (offline wallets) for added security.

The Early Days of Bitcoin and the Birth of Hardware Wallets

Bitcoin, the pioneering cryptocurrency, was introduced in 2009. During its early days, there were no established means for securely storing Bitcoin, especially as the cryptocurrency began to gain traction. Early Bitcoin wallets were clunky and required a level of technical expertise to operate, which made them far from user-friendly. The initial obstacle was the need for users to execute terminal commands, which discouraged widespread adoption.

For those who were early adopters and accumulated Bitcoin during this period, the result was often the loss of funds due to forgotten passwords or corrupted hard drives. It's estimated that of the 21 million Bitcoins ever created, over 3 million were lost in the early years, whether due to sending funds to the wrong address, forgetting private keys, or data corruption. Back then, the value of Bitcoin was relatively low, and many shrugged off the loss. However, as Bitcoin's value increased over time, these early losses became worth millions—if not billions—of dollars by 2022.

The Rise of Bitcoin's Value and the Need for Better Storage Solutions

In 2013, following Bitcoin's first halving event in 2012, the cryptocurrency hit an all-time high of $1,000 per Bitcoin. As Bitcoin's value surged, investors became more concerned about the security of their assets. In response to these concerns, developers began creating innovative solutions to securely store Bitcoin and other cryptocurrencies, sparking the creation of hardware wallets.

Xapo, a company founded in 2013, was one of the earliest companies to provide a secure hardware

wallet solution for cryptocurrency storage. By 2014, Xapo began offering its services to exchanges and early adopters, providing a secure storage solution for the growing number of investors who were increasingly worried about losing their funds.

The Emergence of Consumer-Friendly Hardware Wallets

While Xapo served primarily as an enterprise-grade solution, it wasn't until 2014 that consumergrade hardware wallets began to emerge. The TREZOR, created by SatoshiLabs, was the first widely recognized consumer hardware wallet. TREZOR revolutionized cryptocurrency storage by providing a hardware-based solution that could store private keys offline, away from potential online threats. While other hardware wallets such as the Goochain Citaelle, EliptiBox, and BitLox also emerged around this time, TREZOR gained significant attention due to its usability and security features.

In August 2014, Ledger entered the market with the Ledger Nano, a hardware wallet that became known for its affordable price and advanced security features. One key feature that set the Ledger Nano apart was its two-factor authentication (2FA) card, which added an extra layer of security. Despite the fact that the 2FA card wasn't the most user-friendly option, it was effective in securing private keys and reducing the risk of unauthorized access.

Advances in User Experience and the Popularity of Ledger and KeepKey

While the Ledger Nano was an immediate success, it was clear that further development was needed to improve user experience. In 2015, KeepKey entered the market, offering a hardware wallet with a larger, clearer display. This design enhancement appealed to users who were looking for a more user-friendly experience, helping KeepKey carve out a niche in the market.

In August 2016, Ledger Nano S was released, taking the hardware wallet market by storm. This wallet included the now-standard feature of a screen, improving the user interface and making it easier for users to verify transactions directly on the device. It also continued to include the 2FA card for enhanced security. By the time 2017 arrived, the Ledger Nano S was widely regarded as one of the best hardware wallets on the market.

The Ongoing Evolution of Hardware Wallets

As the cryptocurrency market grew, the need for more sophisticated and secure storage solutions became even more critical. Hardware wallets continued to evolve to address emerging challenges such as scalability, speed, and user experience. Today, hardware wallets remain one of the most secure ways to store cryptocurrency, particularly for long-term holders and investors.

One of the key benefits of hardware wallets is that they store private keys offline, making them highly resistant to hacking attempts that target online wallets. In addition to their security, hardware wallets offer a degree of anonymity by allowing users to interact with cryptocurrencies without revealing personal information. This makes them particularly attractive for privacy-conscious users who value the ability to make secure, anonymous transactions.

The Benefits of Hardware Wallets

The advantages of using hardware wallets cannot be overstated:

- **Offline Storage (Cold Storage)**: Hardware wallets store private keys offline, reducing the risk of hacking and online threats. This makes them one of the most secure storage options for cryptocurrency.
- **Anonymity and Privacy**: Transactions made using hardware wallets are not linked to your personal information, offering an added layer of privacy compared to online wallets.
- **Security**: Hardware wallets are immune to viruses, malware, and phishing attacks that commonly target online wallets.
- **Portability**: Many hardware wallets are small, lightweight, and easily portable, making them convenient for users who need to carry their funds securely.
- **Multi-Currency Support:** Leading hardware wallets support multiple cryptocurrencies, allowing users to store various assets on a single device.

A major benefit of hardware wallets is that they can hold a wide range of cryptocurrencies, including Bitcoin (BTC), Ethereum (ETH), Litecoin (LTC), and many others. As the crypto ecosystem continues to grow and more altcoins are developed, hardware wallets support various tokens and digital assets, offering users a centralized solution to manage their entire portfolio. Whether you're dealing with ERC-20 tokens, NFTs, or assets from new blockchains, a hardware wallet can store them all, reducing the need for multiple wallet applications and simplifying your crypto management.

Why You Should Use a Hardware Wallet for Crypto Security

Cryptocurrency security has been a growing concern for many as the digital asset market continues to expand. Hardware wallets provide an essential level of protection that online wallets (also known as "hot wallets") simply cannot match. In this section, we will continue discussing the key benefits of hardware wallets, including how they keep your funds safe and offer peace of mind when storing cryptocurrency.

Benefits of Using a Hardware Wallet

1. **Private Keys Remain Secure and Offline:** One of the most critical aspects of hardware wallets is that they keep your private keys offline, making them immune to online threats such as hacking. When you use a hardware wallet, the private key remains on the device itself and never gets exposed to the internet, reducing the risk of it being stolen or compromised. This feature is a significant improvement over software wallets (hot wallets), where your private key is stored on your computer or phone, which could be vulnerable to malware or phishing attacks.

2. **Password Encryption and Enhanced Security**: Hardware wallets come with built-in password encryption that protects the device from unauthorized access. Only the user knows the password, and this password is required to access the wallet. If someone tries to force access to your device by entering the wrong password multiple times, the wallet can be set to self-destruct after a specified number of incorrect attempts, ensuring that no one can access your funds. This level of security is not available with online wallets.

3. **Supporting Multiple Cryptocurrencies**: Another major advantage of hardware wallets is their ability to support a wide variety of cryptocurrencies. Some hardware wallets can support up to 10,000 different cryptocurrencies, including popular coins like Bitcoin (BTC), Ethereum (ETH), Ripple (XRP), and Binance Coin (BNB). This makes it convenient to store all your crypto assets in one place, rather than using different wallets for each currency.

4. **Immune to Viruses and Malware**: Since hardware wallets are not connected to the internet, they are immune to viruses, malware, and hacking attempts that often affect software

wallets. Whether your computer or smartphone is infected with malicious software, your hardware wallet remains secure. This is one of the primary reasons why hardware wallets are considered the safest way to store your cryptocurrency.

5. **Transaction Verification on the Hardware Device**: When you make a transaction with a hardware wallet, you are required to verify the transaction directly on the device itself. This adds an additional layer of security, ensuring that no one can make transactions on your behalf without physically accessing your hardware wallet. This is crucial for preventing unauthorized transfers, especially in case your computer or phone becomes compromised.

Essential Tips for Using a Hardware Wallet

Now that you understand the benefits of hardware wallets, it's essential to follow best practices for securing your device and recovery information. Here are four critical safety tips to follow:

1. **Use a Strong, Memorable Password**: The password you create during the setup of your hardware wallet is the first line of defense against unauthorized access. Choose a strong, unique password that cannot be easily guessed. It’s a good idea to write it down and store it in a secure location, such as a safe or a locked drawer. Avoid using easily guessable information like your birthdate or pet’s name. Remember, if you forget your password, you’ll have to reset the wallet using your recovery phrase.

2. **Store Your Recovery Words Safely**: Most hardware wallets generate a recovery phrase (typically 12 or 24 words) when you set up the device. This phrase is the key to recovering your funds in case your device is lost, stolen, or damaged. If someone else gets access to this recovery phrase, they could steal all your cryptocurrency. Therefore, it’s critical to store these recovery words securely, preferably in a fireproof and waterproof container like a bank safety deposit box.

Do not store them digitally (e.g., on your phone or computer), as this could expose you to potential hacks.

3. **Enable PIN Protection**: Most hardware wallets allow you to set up a personal identification number (PIN) that must be entered whenever you access the device. This extra layer of security ensures that even if your hardware wallet is physically stolen, the thief cannot access your funds without the PIN. Choose a PIN that is difficult for others to guess, and make sure to keep it confidential.

4. **Backup Your Recovery Phrase Regularly**: While it's essential to securely store your recovery words, it's also advisable to back them up in multiple safe locations. Having multiple copies of your recovery phrase stored securely can help protect you against loss or theft. Consider storing one copy in a safety deposit box and another in a trusted family member's secure location.

A Must-Have for Crypto Security

The development of hardware wallets has revolutionized the way cryptocurrency holders can secure their funds. By keeping private keys offline, providing enhanced encryption, and offering features like transaction verification on the device itself, hardware wallets offer the highest level of protection available for crypto assets.

If you're serious about safeguarding your cryptocurrency investments, it's time to move your Funds from hot wallets to cold storage with a hardware wallet. By doing so, you'll significantly reduce the risks associated with online threats and ensure your digital assets remain safe from hackers. Additionally, following best practices for setting up and using your hardware wallet will give you peace of mind, knowing that your funds are secure.

In the rapidly evolving world of cryptocurrency, hardware wallets are no longer just a luxury—

they're a necessity. So, take the necessary steps today to secure your assets and protect your digital wealth for the future.

Tips for Securely Storing Your Crypto Funds

1. Split Your Funds Across Multiple Hardware Wallets

One of the most common mistakes is putting all your funds in a single hardware wallet. To mitigate risks, divide your total holdings into two or more hardware wallets based on your preference and comfort. This ensures that even if one wallet is compromised, your entire portfolio is not at risk.

Personally, I use four hardware wallets to diversify and secure my assets effectively.

2. Keep Your Hardware Wallets Secure

Just as you safeguard your physical wallet, you need to protect your hardware wallets from theft or unauthorized access. Since you don't need to carry your hardware wallets daily, store them in a safe yet accessible location. Avoid discussing the contents of your wallet or bragging about your holdings. Discretion is key to security. For added protection, consider keeping your wallet in a hidden or locked area.

3. Stay Informed and Plan for Growth

Selecting the right hardware wallet is an essential step in securing your assets. Conduct your own research (DYOR) to find a wallet that fits your budget and meets your security needs. As your portfolio grows, you can always invest in additional wallets to enhance your security strategy.

By following these tips and choosing a hardware wallet that aligns with your needs, you can ensure your funds remain secure, giving you peace of mind as you grow your portfolio.

Choosing the Best Hardware Wallets for Your Cryptocurrency

When it comes to securing your cryptocurrency, choosing the right hardware wallet is essential. Here are some of the best hardware wallets available today, and why having at least two of them is a smart strategy.

The Two-Wallet Strategy: Trading and Cold Storage

1. **Trading Wallet**: Use one wallet exclusively for trading and transferring funds to exchanges. This wallet is for your active, day-to-day transactions.
2. **Cold Storage Wallet:** Designate another wallet for long-term storage. Treat this as your "retirement account" for cryptocurrency. Funds in this wallet should remain untouched for years, giving you the opportunity to look back and appreciate your foresight. Think of it as a vault for your future wealth.

Top Hardware Wallets Today

Several hardware wallets currently dominate the market, providing users with robust security, ease of use, and excellent customer support. As of the time of writing, here are some of the most popular hardware wallets available:

1. **TREZOR Model T**: The TREZOR Model T is one of the most secure and user-friendly hardware wallets available. It offers an intuitive touchscreen interface and supports a wide range of cryptocurrencies. It is ideal for both beginners and experienced users.
2. **Ledger Nano X**: The Ledger Nano X builds on the success of the Ledger Nano S with Bluetooth connectivity, allowing users to manage their funds through mobile apps. It also

supports over 1,500 cryptocurrencies and offers excellent security features, including a secure chip for storing private keys.

3. **KeepKey**: KeepKey's larger display and simple interface make it a popular choice for those looking for an easy-to-use hardware wallet. KeepKey supports a variety of cryptocurrencies and offers integration with ShapeShift for seamless crypto exchanges.
4. **Trezor One**: The Trezor One is an affordable and secure hardware wallet that supports numerous cryptocurrencies. It is an excellent option for users looking for a no-frills wallet with top-notch security.

Why a Hardware Wallet is Essential

If you're still hesitant, consider the risks. Imagine being the person who lost $3.6 million because they didn't transition their funds to a hardware wallet. That's a cautionary tale you don't want to live. By investing in a hardware wallet, you're not only protecting your funds but also ensuring peace of mind.

Do Your Own Research (DYOR)

There are numerous hardware wallets on the market, each with its unique features and price points. Take the time to research and select a wallet that suits your needs and preferences. As your portfolio grows, you can always expand your collection of hardware wallets for additional security.

Take Action Today

Don't wait until it's too late. Protect your assets now by purchasing a hardware wallet and moving your funds into cold storage. Your future self will thank you for taking this critical step toward financial security.

Chapter 21

Crypto Debit Cards and Credit Cards: Unlocking Power of Rewards

Imagine being able to buy coffee, food, houses, or even cars using cryptocurrency instead of fiat currency. Now, picture earning cashback rewards for these purchases as well. While it might sound farfetched, it's a reality for those who embrace cryptocurrency as a way of life.

From my personal experience, I've purchased cars outright with crypto, paid for houses in Puerto Rico using cryptocurrency, and traveled extensively—all while living off crypto. My journeys have taken me to places like Turks and Caicos, the Bahamas, Egypt, Australia, Germany, Sweden, Greece, the Maldives, and many more. Everything, from gas to medical expenses, has been paid for with cryptocurrency for several years now.

In this guide, I will share how you can integrate cryptocurrency into your daily life using crypto debit and credit cards. We'll also explore some of the best crypto cards currently available, highlighting their features and benefits—which often outshine traditional debit and credit cards.

What are Crypto Debit and Credit Cards?

Crypto debit and credit cards function similarly to traditional cards, enabling users to complete everyday transactions with cryptocurrencies such as Bitcoin (BTC), Ethereum (ETH), XRP, and various altcoins. Backed by major payment networks like Visa and Mastercard, these cards are widely accepted at retailers worldwide, eliminating concerns about merchant acceptance.

Visa recently reported that customers made $2.5 billion in payments using crypto-linked cards in the first fiscal quarter of the year, representing 70% of its crypto transaction volume for the entire previous year. As of now, Visa has partnered with over 65 cryptocurrency platforms—including Coinbase, Binance,

Crypto.com, Gemini, and others—to enable seamless payments with digital assets.

A Brief History of Crypto Debit Cards

The concept of crypto debit cards emerged as a solution to the challenges of spending cryptocurrency for everyday items. Initially, converting cryptocurrency to fiat involved withdrawal fees, exchange fees, and additional hurdles. While Bitcoin ATMs provided some utility, they were not widely available globally. This gap led to the creation of crypto cards, allowing users to bypass conversion complexities and spend crypto directly.

Coinbase was a pioneer in this space, introducing the Shift card. This card enabled users to make purchases directly by deducting funds from their Bitcoin wallets. Since then, numerous crypto debit cards have entered the market, offering expanded functionalities and support for a variety of altcoins.

Crypto Card Features

Like traditional cards, crypto debit and credit cards, link to your digital wallet, enabling you to make

purchases seamlessly. These cards often come with features such as:

- **Cashback Rewards**: Earn rewards for spending cryptocurrency.
- **Wide Acceptance**: Supported by Visa and Mastercard, these cards are accepted by millions of merchants worldwide.
- **Altcoin Support**: Many cards support a range of cryptocurrencies beyond Bitcoin, including Ethereum and lesser-known altcoins.

With Visa's growing partnerships and a global network of 80 million merchants, crypto cards are becoming an essential tool for those looking to live a crypto-powered lifestyle.

Why Choose Crypto Cards?

Crypto debit and credit cards offer:

1. **Ease of Use**: No need to convert cryptocurrency to fiat for everyday purchases.
2. **Global Accessibility**: Spend your crypto anywhere Visa or Mastercard is accepted.
3. **Enhanced Rewards**: Many crypto cards provide rewards and benefits that rival or exceed traditional financial products.

Living a life powered by cryptocurrency is no longer a dream—it's a reality that's accessible to anyone. With crypto debit and credit cards, you can make everyday purchases, earn rewards, and enjoy the convenience of spending digital assets without relying on fiat currency. So, what are you waiting for? Take the first step towards financial independence by exploring the world of crypto cards and integrating them into your life today!

How Do Crypto Debit Cards Work?

A crypto debit card allows you to use your cryptocurrency for everyday purchases, just like a traditional debit card. Here's how it works:

1. **Top Up Your Card**: To get started, you first need to load your crypto debit card with your chosen cryptocurrency. Depending on the card provider, you can fund it with Bitcoin, Ethereum, or other popular cryptocurrencies. You can top up your card through a website or mobile app, making it convenient and accessible.
2. **Ready to Use**: Once the card is loaded, it's ready to use for any purchase. Many crypto debit cards come with higher spending limits and lower transaction fees compared to traditional bankissued cards.
3. **Making a Purchase**: Let's say you buy a family dinner using your crypto debit card. After the transaction, the card issuer converts your cryptocurrency into fiat currency (USD, GBP, EUR,

etc.) to complete the payment. The seller receives the payment in their local currency, while you've paid using digital assets. It's that simple.

4. **Top Up Methods**: You can add funds to your crypto debit card through various methods, such as Apple Pay, Google Pay, ACH transfers, wire transfers, debit or credit cards. You simply purchase the desired amount of cryptocurrency and load it onto the card.

Benefits of Crypto Debit Cards

Now that you know how to load and use a crypto debit card, let's explore the benefits of using one. Crypto debit cards offer numerous advantages. Here are five key benefits:

1. **Liquidity:** Cryptocurrencies are versatile investments, serving both short-term and long-term purposes. You can hold your crypto in a hardware wallet for long-term growth (HODL) or load it onto your crypto debit card for everyday purchases. Whether you're buying breakfast bacon or paying for your morning coffee at Starbucks, crypto debit cards make spending your digital assets simple and efficient.

2. **No Forex Fees:** Traditional bank debit cards often charge high forex fees when you convert one currency into another, especially during international transactions. With most crypto debit cards, you can avoid these excessive fees. Card providers typically charge only standard Visa or MasterCard fees, which are much lower and won't significantly impact your budget.

3. **Low or No Annual Fees:** While many traditional banks charge monthly or annual fees for using a debit card, many crypto debit card providers offer low or even zero annual fees. Some providers may waive these fees entirely if you meet a minimum spending threshold. By using a crypto debit card, you can save on these maintenance fees, making it a more cost-effective option for daily transactions.

4. **Rewards and Incentives:** Some crypto debit cards offer rewards, such as cashback or crypto incentives, for every purchase you make. This can be a great way to earn some rewards while spending your cryptocurrency on regular purchases.
5. **Global Accessibility:** Crypto debit cards are widely accepted and can be used at any merchant that accepts Visa or MasterCard, whether in-store or online. This gives you global access to your cryptocurrency, allowing you to use it almost anywhere, just like a traditional debit card.

In summary, crypto debit cards are a convenient and cost-effective way to use your cryptocurrency for everyday purchases. Whether you want to save on forex fees, avoid annual charges, or enjoy global access to your digital assets, a crypto debit card could be a smart addition to your wallet.

So why wait? Start using your crypto debit card today and unlock the full potential of your digital assets.

One Card, Multiple Currencies to Use

With crypto debit cards available today, you can top up your crypto debit card with many different cryptocurrencies and make payments with whichever cryptocurrency you wish. They also allow you to have both fiat and crypto coins, so you don't have to switch cards.

The Downsides of Crypto Debit Cards

While crypto debit cards offer numerous benefits, it's important to be aware of the potential drawbacks. Here are 3 cons of using crypto debit cards:

1. **Additional Fees:** Although many crypto debit card providers waive fees if you meet certain spending thresholds, there are still several fees to be mindful of. For example, if your card

remains inactive for a long period, or if you load your card with cryptocurrencies that aren't supported by the provider, you could face significant fees. These fees can sometimes be much higher than what you'd pay with a traditional Visa or MasterCard. In addition, there are often transaction fees and other charges that may not apply to standard debit cards, which could add up over time.

2. **Limited Cryptocurrency Options**: While many crypto debit cards support multiple cryptocurrencies, they don't always offer every coin or token you might want to use. If you're a fan of altcoins that aren't supported by your card, you may have to convert your assets into a supported cryptocurrency before loading your card. This could involve additional steps or fees that detract from the convenience of using your crypto for purchases.

3. **Security and Regulatory Risks:** Since crypto debit cards are relatively new, they might not always be as secure or regulated as traditional debit cards. In the event of a security breach or fraud, the recourse available to you may be limited. Furthermore, the regulatory landscape around cryptocurrencies is still evolving, which means that changes in laws or regulations could affect the availability or features of crypto debit cards. Users must stay aware of the potential for sudden shifts in the legal and regulatory environment that could impact their ability to use their cards or access their funds.

While crypto debit cards offer impressive benefits, such as cashback rewards, multiple currency options, and the ability to use crypto in everyday transactions, it's important to consider the potential fees and other limitations. Weighing the pros and cons can help you make an informed decision about whether a crypto debit card is the right choice for your financial needs.

Top Crypto Debit Cards to Consider in 2025

Cryptocurrency debit cards are becoming increasingly popular, offering a seamless way to spend your crypto assets in the real world. Whether you want to earn rewards, avoid high fees, or simply use your digital assets for everyday purchases, there are plenty of options to choose from. Below, we'll explore some of the most popular crypto debit cards, including Crypto.com, Coinbase, Wirex, and Nexo, along with a few other notable cards.

1. **Crypto.com Debit Card:** The Crypto.com Debit Card is known for its up to 8% cashback on purchases, making it one of the most rewarding crypto debit cards available. However, it comes with a tiered system, with five different card levels that cater to different users. To unlock the best benefits, you'll need to stake CRO (Crypto.com's native cryptocurrency), which can be a significant upfront investment. The higher the tier, the more rewards and exclusive perks you can access.

Card Perks (Premium Tiers):

- 100% cashback on Netflix, Spotify, and Amazon Prime subscriptions
- 10% cashback on Expedia and Airbnb bookings
- Airport Lounge Access for you + 1 guest
- Private Jet Partnerships
- Crypto.com Private membership
- Exclusive Merchandise Welcome Pack
- Bonus rewards for spending on selected categories

The card supports over 100 cryptocurrencies and allows you to earn rewards in CRO for every purchase. Depending on your tier, fees, and staking levels vary, with some cards requiring a stake of up to $400,000 for the Obsidian Black tier. For beginners, the Midnight Blue card has no staking requirement, but the perks are more limited.

2. **Coinbase Visa Debit Card:** The Coinbase Visa Debit Card is a solid choice for those who want a simple, low-fee crypto debit card with the added benefit of Bitcoin rewards. Coinbase offers zero annual fees and allows users to earn 1% Bitcoin rewards or 4% Stellar Lumens rewards. The card supports a wide range of cryptocurrencies, including Bitcoin, Ethereum, Bitcoin Cash, Ripple, Stellar Lumens, and Litecoin.

   Key Features:

   - International Purchases with no foreign transaction fees
   - Contactless Payments and integration with Apple Pay and Google Pay
   - ATM Withdrawals worldwide
   - Cashback on all purchases
   - Zero sign-up fees and a daily limit of $2,500 for purchases and $1,000 for ATM withdrawals

However, one downside is the 2.49% transaction fee that applies to each purchase or ATM withdrawal. While the fee is relatively high compared to some competitors, Coinbase offers a secure and easy-to-use mobile app to track your spending and rewards.

3. **Wirex Debit Card:** Wirex is a well-known name in the crypto card space, offering a 2% cashback on purchases made with the card. What makes Wirex stand out is its wide acceptance of 36 cryptocurrencies, which gives you more flexibility in how you use your

card. Wirex also makes it easy to convert crypto into traditional fiat and spend it at over 61 million merchants worldwide.

Key Features:

- 2% Cashback on every purchase
- Supports 36 cryptocurrencies including Bitcoin, Ethereum, Litecoin, and more
- SG $1,400 daily ATM withdrawal limit
- Easy-to-use mobile app for managing funds and converting between crypto and fiat

Wirex does not charge an application fee and offers a no-fee option for certain activities, though transaction fees and conversion fees may apply. It's a good option for users looking to diversify the cryptocurrencies they hold while still getting a competitive cashback rate.

4. **Nexo Debit Card:** The Nexo Debit Card is unique because it allows you to use your crypto assets without needing to sell them. Instead, you can borrow against your crypto holdings, which means you avoid triggering a taxable event with the IRS. This makes it a tax-efficient solution for crypto enthusiasts. Additionally, it's issued as a Mastercard, meaning it's accepted by over 40 million merchants worldwide.

Key Features:

- No need to sell your crypto assets—borrow against your crypto holdings instead
- Global Acceptance at 40 million merchants via Mastercard
- Nexo Wallet App for managing your card and crypto balance
- Earn rewards on every purchase made with the card

With the Nexo Wallet App, you can easily track your spending, freeze your card if lost, or use a virtual card for online purchases. The Nexo card also doesn't charge any hidden fees, which makes it a good choice for frequent users.

Overall, it is a good card. The crypto debit card supports 45 fiat cryptocurrencies. The amount of loan a cardholder can get depends on how much they hold in their wallet. This is very beneficial for people who want to continue to hold their cryptos until they purchase something with them. The interest rate for the card is 5.9%, and there are no exchange fees. When cardholders make a purchase, they get 2% instant cash back in either BTC or Nexo tokens. Nexo has partnered with MasterCard so cardholders can use the Nexo card anywhere MasterCard is accepted.

5. **Revolut Crypto Debit Card:** Revolut is a well-known fintech platform offering a crypto debit card that allows users to buy, sell, and spend cryptocurrency. The Revolut card supports over 30 cryptocurrencies, including Bitcoin, Ethereum, and Litecoin. Revolut users can also access a range of traditional banking services like global payments, money transfers, and savings accounts.

Key Features:

- Spend up to 30 cryptocurrencies with ease
- Crypto-to-fiat conversion at no extra charge
- Available in multiple currencies, with no foreign exchange fees for premium users
- Instant spending notifications through the mobile app

Revolut is ideal for users looking for an all-in-one financial solution with the added benefit of using crypto for everyday purchases. However, there are some fees for higher-tier users, particularly those who want to access premium benefits like free international ATM withdrawals or higher spending

limits.

6. **BitPay Card:** The BitPay Card allows users to convert and spend Bitcoin and Bitcoin Cash at over 50 million merchants globally. It's linked to your BitPay account, where you can convert your crypto into fiat before using it for purchases. The card doesn't charge an annual fee but does come with a small setup fee.

Key Features:

- Spend Bitcoin or Bitcoin Cash easily
- No annual fee (though a setup fee applies)
- Cashback rewards available in crypto
- Free ATM withdrawals (within certain limits)
- Instant crypto-to-fiat conversion

It supports eight fiat currencies and the following crypto:

- Bitcoin
- Ethereum
- Gemini Dollar
- USD Coin
- Bitcoin Cash
- Paxos

There are daily spending limits of $10,000 and the cardholder can hold a maximum of $25,000 in their account. The card has a 3% foreign transaction fee, a $2.50 fee for ATM withdrawals, and a $5 monthly inactivity fee if a cardholder doesn't use their card for 90 days. The BitPay card has users pay a one-time activation fee of $9.95, and individuals must verify their identity. Cardholders can add cryptos into their BitPay wallet and from there convert them into fiat currency and add that to their card.

The BitPay card is a great choice for Bitcoin holders who want to avoid selling their assets and still enjoy spending them at everyday retailers.

7. **BlockCard Debit Card:** Currently, this card is for US residents only and will soon be available for several European countries as well. With a cashback reward of up to 6%, its popularity among users is high. This crypto card charges no deposit fees, no annual fee, or no exchange fee. There are different types of block cards available that offer different levels of operations. The best thing is that there is no limit on cashback after a certain amount of spending. As a result, users can spend any amount of money and receive unlimited cashback. If you spend $750 every month, you do not need to pay its monthly fee. ATM transactions are a bit costly at $3 per withdrawal. It is a great card if you know how to use it.

    Key Features:

    - No deposit or exchange fees
    - $10 activation fee
    - $3 fee for ATM withdrawals
    - $5 monthly fee, waived for cardholders who spend $750 per month.

BlockCard supports several cryptocurrencies, and it works with Google Pay and Apple Pay. Staking is required for all cardholders.

8. **CoinZoom Debit Card:** CoinZoom offers a Visa crypto debit card that supports more than four dozen cryptocurrencies. There are a few different cards to choose from, all of which require staking the ZOOM token to get rewards and cashbacks. Cardholders can easily switch between spending different cryptocurrencies in their CoinZoom wallet. Conversions do not occur until the cardholder makes a purchase. There is a $10 activation fee required and then $0.30 per transaction after that.Cardholders also get access to ZoomMe, CoinZoom's peer-to-peer payments platform.
9. **Swipe Visa Debit Card:** The Swipe Visa debit card is a dual interface NFC/EMV physical card and three of the four can do ATM cash withdrawals. That means users can choose from among four optional card tiers with different percentages of rewards like Saffron, Sky, Steel, and Slate.

   Users can get cashback by shopping on Spotify, Netflix, Apple Music, Hulu, and Amazon Prime. Get extra cashback rewards for Travala, Uber, Starbucks, and Airbnb. The entry level for this crypto debit card costs $25, and there is no requirement to stake the tokens. You get 1% cashback paid in Bitcoins. Sky offers 2% cash backs and is a must to stake 300 SXP. With the Steel option, you get 4% cash backs and must stake 30,000 SXP to get the card. Users can earn more rewards by spending SXP tokens, which can be used on Ethereum and Binance Chain. Users can stake the token to earn staking rewards.

   Key Features:

   - Integration with Apple Pay, Samsung Pay, and Google Pay to allow users to pay for goods and services with the card.

- Get 1-5% Cashback. 10-100% on specific purchases.
- No option to load with fiat.
- Supports seven cryptocurrencies that can be loaded for spending with the card.
- When you stake tokens to get the cards, the lock-up period is 6 months.

Fees: Steel and Slate users do not pay foreign transaction fees, others do. $25-$50 to get cards in addition to staking costs. You must stake the platform tokens to get some card tiers. 2% foreign transaction exchange fees apply.

10. **Gemini Crypto Credit Card:** Gemini Crypto Credit Card is available to all users worldwide and connected to MasterCard. Unlike other crypto cards that prevent you from accessing realtime rewards for purchases, Gemini offers real-time rewards on purchases for Bitcoin or other cryptos on Gemini. You get up to 3% cashback on dining purchases, 2% on groceries, and 1% on other purchases. It will be possible to load with 30 cryptocurrencies available on Gemini. The card issuer is WebBank, a digital lending bank.

Key Features:

- Instant up to 3% cash backs deposited into the user's Gemini account.
- Has an iOS and Android app to accompany and manage the card. Can freeze the card instantly from the app, track expenses, etc.
- In-app access of your credit card number and Apple Pay or Google Pay after approval. Residents in all 50 U.S. states can now apply to join the waitlist. You do not need to be a Gemini user to apply but will need a Gemini crypto exchange account where the rewards will be deposited.

- Black, silver, and rose gold metal cards.
- Provides 24/7 live agent support. Near-instant access to the card after application.

Fees: There is no fees to exchange and no cost for acquiring the card. No annual fee.

Choosing the right crypto debit card depends on your specific needs—whether you want high cashback rewards, low fees, the ability to use multiple cryptocurrencies, or even earn Bitcoin on your purchases. Cards like Crypto.com and Nexo provide premium perks, while options like Coinbase and Wirex offer flexibility and ease of use. Whether you're a frequent spender, a traveler, or someone looking to make the most of your crypto assets, there's a card that's perfect for you.

By considering factors like fees, supported cryptocurrencies, rewards programs, and card benefits, you can make an informed decision on which crypto debit card is the best fit for your financial goals

Why Crypto Debit and Credit Cards Are a Game-Changer

As you can see, there are numerous benefits to using crypto debit and credit cards, with plenty of options to choose from. Crypto cards offer you the freedom to live life on your own terms—whether it's buying a car, purchasing a home, or even taking your family on a trip to Paris. What's even more remarkable is that many of these cards come with travel insurance and medical coverage, offering a range of benefits that traditional debit cards simply can't match.

Why Crypto Debit Cards Stand Out

One of the key advantages of crypto cards is their versatility. Some cards offer 100% reimbursement on popular subscriptions like Netflix, Amazon Prime, and Spotify, while others give you 10% cashback on bookings through Airbnb and Expedia. For those who enjoy the finer things in life, some premium crypto cards even offer exclusive access to private jet partnerships and other luxury perks. These benefits aren't just for the elite; they're designed to help anyone maximize their spending and rewards.

Privacy and Security

Another important feature is the privacy that crypto cards offer. Unlike traditional debit and credit cards, which are tracked by banks and financial institutions, crypto cards allow for more discreet transactions. This means your spending habits aren't being monitored, providing a level of financial freedom that is often difficult to achieve with traditional banking systems.

Maximize Your Benefits

If you're someone who wants to manage your finances in a more flexible and rewarding way, crypto debit cards are a great option. In fact, you may want to consider holding multiple crypto debit cards, as each one offers unique perks and rewards. By having a few different cards, you can tailor your spending to get the best benefits in various areas, from cashback to luxury services.

Consider Local Regulations

However, before you start collecting multiple crypto cards, make sure to check the regulations in your country. Not all countries allow access to certain crypto debit or credit cards due to local laws or financial regulations. It's essential to verify that the cards you're interested in are available in your region.

Break Away from Traditional Banking

It's time to break free from the confines of centralized banking systems. With crypto debit and credit cards, you can take control of your financial future and become your own decentralized bank. So, what are you waiting for? Start exploring the world of crypto cards today and take full advantage of the rewards, privacy, and benefits they offer.

Chapter 22

The Keys to Profits:

Market Research, Fundamentals, Sentiment Analysis and Technical Analysis

One of my favorite quotes from motivational speaker Eric Thomas is, *"When you want to succeed as bad as you want to breathe, then you will be successful."* This quote resonates deeply with me because it highlights how crucial desire and persistence are in achieving success. It underscores the importance of striving for your goals with the same urgency and determination as the body's need for air.

Breathing is essential to life. Without oxygen, our bodies can survive only for a few minutes. Studies show that even just 30 to 180 seconds of oxygen deprivation can cause significant consequences—loss of consciousness, brain cell damage, and in extreme cases, death. After just one minute, brain cells begin to die, and after three minutes, the damage to neurons becomes irreversible. At five minutes, the body reaches a critical point where survival becomes unlikely.

While this may sound intense, it's an analogy to illustrate the level of urgency you must have in pursuing your success. Just as the body craves oxygen, you must crave success. Not just for yourself, but for your family and those who depend on you. Every failure you encounter, every setback you face, should be met with the same persistence you would have in a life-or-death situation. There is no "I'll start tomorrow." Success doesn't come to those who procrastinate; it comes to those who commit fully and take action.

The First Step: Commitment

The first step in achieving success, particularly in the cryptocurrency space, is making a commitment to yourself and to your family. Commitment is the foundation of success. Without it, your dreams remain distant and unachievable. If you can't make that commitment today—right now—then your dreams will always stay out of reach.

In this book, we've covered a wealth of information to help you navigate the cryptocurrency landscape. The goal is not just to help you make profits, but also to position you for long-term success in a space that's still evolving. There's a lot more to cryptocurrency than just Bitcoin and blockchain. In this chapter, I'll be sharing key insights that will help you continue to succeed in this field, both financially and personally.

A Personal Mission

The primary reason I wrote this book is to empower individuals—just like you—to thrive in the cryptocurrency world. I want to help you not only succeed but also pass this knowledge on to your loved ones. There is a lot of information out there, but not many resources dive as deep into the broader crypto ecosystem as I've aimed to in these pages. Many books focus narrowly on Bitcoin and blockchain technology, but the world of cryptocurrency extends far beyond that, offering opportunities in areas like altcoins, DeFi (Decentralized Finance), NFTs, and more.

This book is designed to give you the tools and knowledge you need to succeed in this space, and most importantly, help your family succeed with you. The cryptocurrency world is vast, and with the right guidance, you can tap into opportunities that were previously out of reach.

The Importance of Market Research in Cryptocurrency

The world of cryptocurrency is vast, but because the internet is so centralized and controlled by private companies, finding reliable and accurate information can be challenging. It often requires intense research or tapping into crypto communities, which can be time-consuming. This is one of the main reasons I decided to dedicate my time to not only creating a crypto consulting business but also writing this book— to provide you with a valuable tool that will give you the confidence to know where to look and how to succeed in the crypto space.

In this chapter, I'll be sharing insights on how to conduct proper market research. You'll learn how to identify the best cryptocurrency websites for research, allowing you to find promising projects early and potentially profit from them. I'll also explain the key analytical tools—Fundamental Analysis, Sentiment Analysis, and Technical Analysis—and show you how to use these tools to make smarter, more informed decisions when evaluating a cryptocurrency or a project. These strategies will help you avoid the common pitfall of rushing into a project without doing thorough research, which can lead to significant financial losses.

This book is your guidebook to the crypto space. Keep it by your side at all times, because it's not just a book—it's your mentor and your best friend in this journey. There's a lot of valuable information packed in these pages, and I strongly advise you to read it more than once. If you skim through it or only read it once, you'll miss out on crucial details. I encourage you to revisit the material as often as you can— take notes, highlight key sections, and mark this book up. Treat it as your instruction manual for navigating the cryptocurrency market.

The Hidden Gems in This Book

This book is more than just a collection of tips and strategies—it's a treasure chest of knowledge that could change your financial future. The value of what you've purchased goes far beyond the price you paid for it. If you take this information seriously and apply it diligently, it can help you achieve financial freedom.

Remember, the first step towards financial freedom is making a commitment. The second is to take massive action, and the third is to stay curious—always learning something new every day.

Market Research: What It Is and Why It Matters

Now, let's dive into market research. In the cryptocurrency world, market research refers to the process of assessing the potential of a cryptocurrency or blockchain project. It involves evaluating the viability of the project by gathering data from the community, target market, and other relevant sources.

Market research allows you to discover the target audience for a cryptocurrency and gauge the level of interest and feedback from the community. It gives you insights into the project's value and its chances of success. However, it's important to exercise patience during this process. Rushing through market research can lead to poor decisions and financial losses. Proper due diligence is essential to protect your investments.

The Power of Market Research

Effective market research can position you to enter a cryptocurrency project early, allowing you to maximize profits. For example, imagine you do thorough research and find a cryptocurrency that solves a real-world problem, has a strong community backing, and is priced at just $0.03. If the project continues to gain traction and the price soars to $16 in a few weeks, your early investment would yield massive profits. This is the power of good market research—it can make the difference between making a fortune or losing everything.

That's why market research is not something to be taken lightly. A single mistake can cost you your entire investment, so be sure never to rely on just one source of information. Blindly buying a coin or entering a trade based on limited data can lead to disastrous results. The crypto market is volatile and unpredictable, and without solid research, your trading account could be wiped out.

The Three Key Tools for Effective Market Research

To minimize losses and increase your chances of success, there are three key methods you should use to conduct market research in cryptocurrency: Fundamental Analysis, Sentiment Analysis, and Technical Analysis. Let's start with Fundamental Analysis (FA), the most important tool in your research toolkit.

Fundamental Analysis (FA)

Fundamental analysis is the first step in evaluating the potential value of a cryptocurrency project. It helps you understand the utility of a coin or token—essentially, what problem it's solving and how it's improving the market. Before investing, make sure the cryptocurrency project you're considering has a realworld use case and tangible value. Ask yourself: Does this project offer something that current markets need? Does it solve a problem or improve an existing solution?

If the project lacks a solid use case or utility, it's a red flag. Investing in such projects is often a recipe for failure. By focusing on projects that add real value to society, you can eliminate those that are more speculative or lack long-term potential. This approach helps you save time, reduce frustration, and avoid unnecessary losses.

Understanding Fundamental Analysis in Cryptocurrency

Fundamental analysis (FA) plays a crucial role in cryptocurrency investing. It involves evaluating the intrinsic value of a cryptocurrency by examining a variety of factors, such as the project's team, partnerships, whitepapers, roadmaps, and more. The primary goal of fundamental analysis is to determine whether the current price of a cryptocurrency reflects its true value based on these underlying factors. With this insight, you can make more informed, long-term investment decisions while minimizing risk.

It's important to note that fundamental analysis does not rely on past market performance to predict future price action. Instead, it focuses on macroeconomic and microeconomic factors that could influence the success or failure of a cryptocurrency or project in the future. By evaluating these factors, you can better gauge whether a cryptocurrency has the potential for long-term growth or if it's just another speculative asset.

Key Components of Fundamental Analysis

When performing fundamental analysis, there are several critical elements to evaluate:

1. **The Team**: One of the most significant indicators of a cryptocurrency project's potential is the background of its team. Research the founders and the board of directors—do they have substantial experience in the cryptocurrency and blockchain industry? Have they worked on any other successful projects? More importantly, do they have a history of successful exits and have they been involved in any past scams? A team with a proven track record of success provides confidence in the project's future.

2. **The Whitepaper**: The whitepaper is the foundation of every cryptocurrency project. It outlines the technical and business aspects of the project, including its goals, utility, and the problem it aims to solve. A solid whitepaper provides clear and detailed information on how the project will succeed. Make sure to thoroughly review it to assess the feasibility and ambition of the project.

3. **Roadmap**: The roadmap details the project's planned milestones and future development. Does the project have a clear, realistic timeline for achieving its goals? A well-defined roadmap can give you a sense of the project's commitment to growth and innovation. Make sure to evaluate whether the project is on track with its goals and timelines.

Market Conditions and Hype

Another factor that influences cryptocurrency prices is the overall market sentiment and hype. For example, if the market is in a bullish phase, the prices of many cryptocurrencies may rise simply because investor sentiment is positive. A recent example of this was ApeCoin—within three days of its launch, the project attracted over $5 billion in investments. This kind of hype can propel a cryptocurrency's value to new heights, but it's important to remember that hype is often temporary.

In many cases, when Bitcoin experiences a price surge, other altcoins tend to follow. Bitcoin's dominance drives the market, and when it pumps, investor money often flows into altcoins. However, once Bitcoin's price starts to consolidate or dip, profits from the pump tend to trickle into altcoins, causing them to rise as well. This phenomenon is a key aspect of Bitcoin's market dominance and how it influences the broader cryptocurrency market.

The Impact of Bear Markets

While bull markets drive most cryptocurrencies up, bear markets can be a different story. During downtrends, many "paper hands" (those who sell their holdings at the first sign of a market drop) quickly exit the market, leaving behind the "diamond hands"—long-term investors who hold onto their assets despite market volatility. These long-term holders believe in the fundamental value of a cryptocurrency and its use case, and they are often the ones who stabilize the market during downturns by buying the dip when others are selling in fear.

Understanding this market behavior is crucial. While short-term traders may chase profits based on hype, diamond hands are the ones who drive the long-term value of a cryptocurrency. These holders are vital to the survival and recovery of a project, especially when the market is in a downtrend.

Why You Should Perform Fundamental Analysis

Understanding and applying fundamental analysis is essential for several reasons:

1. **Risk Management**: Fundamental analysis helps you identify high-quality projects with strong use cases, a solid team, and a clear vision. This minimizes the risk of investing in low-value or "pump-and-dump" projects.

2. **Long-Term Profitability**: By evaluating the intrinsic value of a cryptocurrency, you can invest in projects that have real potential for growth, rather than chasing hype-driven assets. This approach increases your chances of long-term success.

3. **Market Insight:** Performing fundamental analysis helps you understand how market conditions—such as bullish or bearish sentiment—affect the price action of a cryptocurrency. This insight enables you to make smarter decisions based on real data rather than speculative trends.

By consistently applying fundamental analysis to your investment decisions, you'll be better equipped to navigate the complex world of cryptocurrency and make more informed, profitable choices.

How Fundamental Analysis Can Guide Your Investment Decisions

Fundamental analysis isn't just about identifying undervalued cryptocurrencies; it also plays a crucial role in shaping your trading and investment strategy. Here are three key ways that fundamental analysis can help you make smarter decisions in the crypto market:

1. **Making Better Mid-Term Trading Decisions in Bearish Market Conditions:** When the market is in a downtrend, also known as a bear market, it's easy to feel overwhelmed by price drops. However, by applying fundamental analysis, you can make more informed decisions. It helps you identify projects that have strong long-term potential, even in challenging market conditions. This knowledge can give you the confidence to hold or accumulate assets while others panic, leading to better opportunities when the market rebounds.

2. **Building Confidence for Long-Term Investments:** Fundamental analysis gives you the information you need to be more confident in your long-term investments. When you understand a cryptocurrency's true value, its team, mission, and the problem it solves, it becomes easier to make decisions based on logic rather than short-term price swings. This sense of confidence can help you buy and hold cryptocurrencies for longer periods without second-guessing your choices.

3. **Identifying Real Innovation vs. Copycats or Scams:** In the rapidly evolving world of cryptocurrency, not all projects are worth your investment. Fundamental analysis helps you spot the real innovators—those projects that solve actual problems and drive the space forward. On the other hand, it also helps you avoid projects that are mere copycats or potential scams. This is especially important when looking to invest in Initial Exchange Offerings (IEOs) or other earlystage projects. Conducting thorough research can save you from falling for projects that promise big returns but lack the fundamentals to succeed.

Step-by-Step Guide to Conducting Fundamental Analysis

Now that you understand the importance of fundamental analysis, let's dive into a simple, effective process you can follow to assess any cryptocurrency. Whether you're analyzing an established cryptocurrency or exploring a new project that piques your interest, this approach will guide you through the key steps.

Step 1: Find a Cryptocurrency That Interests You: The first step is to identify a cryptocurrency that you're interested in learning more about. You can do this by visiting popular cryptocurrency data platforms like CoinMarketCap, CoinGecko, CoinPaprika, or CoinCheckup. These platforms offer comprehensive data on thousands of cryptocurrencies.

Step 2: Check Out the Project's Website and Whitepaper: After identifying a cryptocurrency, the next crucial step is to investigate the project's official website. This is where you can find information about the project's concepts, its utility, and how it aims to solve current issues within its target marketplace. A legitimate project will have a clear, informative website that lays out its mission and how it plans to execute its goals.

Along with the website, make sure the project has technical documentation available. This documentation typically comes in the form of a whitepaper or litepaper. The whitepaper is essential for understanding the technical aspects of the project—how it operates, what technology it's built on, and what problems it intends to solve. For example, when researching Terra (LUNA), you would look for a detailed whitepaper that explains the project's functionality, tokenomics, and roadmap.

Step 3: Check Social Media Platforms: Next, dive into the community's sentiment around the cryptocurrency. Social media platforms like Twitter, Discord, Telegram, and YouTube are invaluable tools for gauging community interest and feedback. In these communities, users are often very vocal about their opinions—whether positive or negative.

- On Twitter, check the official accounts for updates, and see how the community is responding.
- On Discord or Telegram, browse through the channels to see how active the community is. Are people discussing updates or expressing concerns?
- On YouTube, look for reviews or reactions to the project to get an external opinion.

A strong, active community can be a good indicator of a project's legitimacy and long-term potential. If the community is highly engaged and supportive, it's a positive sign. If there's a lot of negative sentiment or unsubstantiated rumors, it could be a red flag.

Step 4: Check Blockchain Transactions: Blockchain explorers are powerful tools that let you see real-time transaction activity. For example, you can visit Etherscan.io (for Ethereum-based projects) to monitor the movement of a cryptocurrency. By looking at transaction data, you can see how frequently the project is being traded.

- If you find that there hasn't been any notable transaction activity for several days or weeks, the project might be inactive and could be considered a dead project. This lack of activity is a major red flag.
- On the other hand, frequent transactions and increasing trading volume can indicate healthy investor interest and an active market.

Step 5: Check Historical Data: Historical data is another key factor in your research. Platforms like

CoinMarketCap or CoinGecko provide comprehensive charts showing cryptocurrency All-Time High (ATH) and All-Time Low (ATL).

- All-Time High: If the ATH was recently set (e.g., a few hours ago), be cautious. It could mean you're buying in at a market peak, and prices may be due for a correction.

- All-Time Low: Similarly, the ATL can provide insight into whether the coin has already suffered a significant drop, which might signal that it has bottomed out and could be ready for recovery.

By looking at the price history, you can assess the market cycles and avoid buying in at the top of a bull run. This insight helps you plan your entry and exit points more effectively.

Step 6: Check Out the Blogs: Blogs and online publications, like those on Medium or Reddit, are excellent sources for ongoing project updates. Many crypto projects use blogs to announce partnerships, new developments, and progress reports.

- If the project has a blog page that is regularly updated, it shows that the team is committed to communicating with the community. This transparency is a good sign of a legitimate project.
- If a project doesn't have a blog or stops updating it regularly, this could be a major red flag. A lack of updates may indicate that the project is either stagnant or possibly even a scam.

Step 7: Investigate the CEO and Developers: The team behind a cryptocurrency project is one of the most important factors to assess. A project is only as strong as the people behind it. To evaluate the credibility of the team, start by researching the CEO and key developers:

- Check their LinkedIn profiles to verify their backgrounds. Have they worked on reputable projects before? Do they have a history of success, or have they been involved in scams or failed ventures?
- Search for interviews or articles that mention the team members. Are they well-known figures in the cryptocurrency space? Do they have relevant experience?

The track record of the team can tell you a lot about the likelihood of the project's success. If the developers have experience with successful crypto projects, that's a positive indicator. But if they have a questionable history, it could be a huge red flag.

Step 8: Check the Markets the Cryptocurrency is Listed On: Next, it's important to check which exchanges the cryptocurrency is listed on. Cryptocurrencies listed on reputable exchanges like Binance, Crypto.com, Coinbase, Gemini, and FTX have already passed rigorous listing requirements. These exchanges typically only list projects that meet high regulatory standards and have shown strong fundamentals.

If a cryptocurrency is listed on major exchanges, this is usually a good sign that the project is credible. If it's only found on obscure exchanges or decentralized platforms with low liquidity, this could be a red flag. Major exchanges are also quick to delist projects that engage in unethical practices, such as exit scams or rug pulls.

Step 9: Check How the Community Feels About the Project: The sentiment of the community around a cryptocurrency or project can tell you a lot about its potential success or failure. Communities are usually vocal about their thoughts, both positive and negative. If the community on platforms like Twitter, Telegram, Reddit, or Discord seems excited and engaged, it may be a good sign that the project has a solid following.

However, if the community's mood is mostly negative, full of complaints, or there's significant discontent about the project's direction, you should be cautious. Negative feedback could point to issues such as a lack of transparency, failed promises, or underperformance, which might eventually lead to the project's downfall. On the other hand, if people are enthusiastic, sharing success stories, and actively participating in discussions, this is usually a good indicator that the project has solid prospects.

Step 10: Look at Third-Party Data to Validate the Credibility of the Project: Beyond social media, it's important to validate a cryptocurrency credibility by looking at third-party sources. Articles, news outlets, and independent reviews from credible analysts or cryptocurrency influencers can offer valuable insights into a project standing in the market. Websites such as CoinTelegraph, Decrypt, or Bitcoinist can provide trustworthy news and updates on specific projects.

Also, look for peer-reviewed analysis or articles that discuss the project's technicals, economics, or the problem it's aiming to solve. Be wary of projects that only have positive reviews or are being promoted without any real critique—this could indicate manipulation or hype.

Step 11: Check the Number of Circulating and Total Coins: Understanding the circulating supply and total supply of a cryptocurrency is crucial in assessing its potential for price appreciation or depreciation. A project with a low circulating supply and a capped total supply (such as Bitcoin with its 21 million cap) tends to create scarcity, which can drive demand and thus the price up over time.

In contrast, a cryptocurrency with an inflated supply, like Shiba Inu, may struggle to appreciate in value. For example, if a coin has a total supply in the trillions, it would take an extremely high market capitalization for it to reach a significant price. This makes these types of projects much less likely to produce the same kind of gains that low-supply coins might see.

Check the project's whitepaper or coin listing on sites like CoinMarketCap or CoinGecko to learn about their supply schedule. Be wary of projects with no clear details on their circulating supply or inflation model.

Step 12: Check What People Are Buying in Real-Time Around the World: You can gain valuable insights into what's trending in the global market by checking sites like FiatLeak.com, which tracks the global flow of cryptocurrency transactions in real time. By observing what people are buying in different countries, you can get an idea of what's gaining traction and what might be undervalued or underappreciated at the moment.

This is particularly useful if you're looking for emerging trends or projects that are gaining adoption. If a cryptocurrency has a large volume of transactions from countries with emerging markets or techforward communities, it could indicate growing demand and long-term viability.

Fundamental Analysis Is Just the Beginning

Once you've completed this fundamental analysis, you'll have a much better idea of whether a cryptocurrency or project is worth investing in. However, remember that fundamental analysis is just the first step in your journey. To make more informed decisions, you will also need to dive into sentiment analysis, technical analysis, and market trends.

By following these steps and consistently doing your own research, you can significantly increase your chances of finding solid investment opportunities and avoiding scams or poorly executed projects. And as always, ensure that you have a risk management strategy in place and never invest more than you can afford to lose.

Sentiment Analysis: Understanding Market Hype and Trends

In the world of cryptocurrency, sentiment analysis is a powerful tool to gauge the mood of the market—specifically, how traders and investors feel about a particular cryptocurrency or project. It's all about understanding the "buzz," the trends, and the demand surrounding a token, as these factors can have a direct influence on its price movement. Think of sentiment as the emotional pulse of the market, and by analyzing it, you can better predict price fluctuations and make more informed decisions.

Understanding Sentiment Analysis

To put it simply, sentiment analysis helps you tap into the collective energy of the market. You're looking to see how much excitement, optimism, or fear surrounds a specific cryptocurrency or project. If a coin is buzzing with positive sentiment, it may indicate that it's gaining traction, while a drop in sentiment might indicate that the project is losing steam.

Example to Illustrate: Take a look at ApeCoin (APE) as a recent case study. Within just three days of its launch, the project saw a 24-hour trading volume exceeding $3 billion and was listed on major exchanges almost immediately. Why? The intense hype and demand around it. This type of momentum, driven by sentiment, can significantly impact a coin's marketcap and price.

However, it's important to recognize that sentiment doesn't always lead to long-term success. While bullish sentiment can drive up prices, it doesn't guarantee sustained growth. In fact, sometimes a euphoric sentiment could signal the start of a bear market. This happened in 2017 during the cryptocurrency boom and again in 2021 when the market was flooded with bullish sentiment, only for many projects to experience significant pullbacks soon after.

How Sentiment Analysis Can Help You

Sentiment analysis is a tool that can guide your decision-making by providing insights into market cycles and trends. Here's how it can help you:

1. **Predict Mid-Term Price Performance:** When there's a surge in public interest and hype around a cryptocurrency, it might indicate that the coin is nearing its All-Time High (ATH).

Knowing this can help you decide when to sell or take profits before a possible market correction occurs. By reading the crowd's sentiment, you can gauge the peak and dip phases.

2. **Make Decisions Based on Logic Instead of Emotions:** Sentiment analysis helps you navigate the fear of missing out (FOMO) and fear, uncertainty, and doubt (FUD). These emotional triggers often cloud your judgment and cause impulsive decisions. With sentiment data at hand, you can make rational, data-backed decisions. For example, even if everyone around you is frantically buying a trending coin, sentiment analysis can help you determine whether it's a solid long-term investment or just a short-term hype.

3. **Capitalize on Trends:** Sentiment helps you track trends. In many cases, retail investors follow the "crowd mentality," buying into projects that are generating buzz. By analyzing the market mood, you can spot upcoming trends and get in early before the mainstream rushes in, maximizing your potential ROI. If you notice positive sentiment and a growing community around a project, it may be a sign to buy early and ride the next wave.

4. **Make Long-Term Bets:** Beyond just riding short-term waves, sentiment analysis can help you assess whether a project is undervalued and has the potential for long-term growth. If a cryptocurrency has strong fundamentals and a growing, loyal community, its future demand could increase—leading to higher valuation over time. By keeping an eye on sentiment, you can identify hidden gems in the market before they skyrocket.

Where to Find Sentiment Data

To perform effective sentiment analysis, you need to explore various social media channels, forums, and communication platforms. These are where the community of traders and investors share their opinions, thoughts, and emotions about a specific cryptocurrency. Understanding the general sentiment in these communities will give you valuable insights into a cryptocurrency's potential.

Platforms to Check:

- **Twitter**: The hashtag culture here is huge. You can search for tags like $XRP or $BTC to see what the community is saying. Pay attention to trending topics and discussions that revolve around certain coins.
- **Reddit**: Subreddits like r/CryptoCurrency or specific token-based subreddits (e.g., r/Bitcoin, r/Ethereum) are goldmines for sentiment analysis. Reddit users are often vocal about their experiences with projects, both good and bad.
- **Telegram/Discord**: These chat platforms are home to active cryptocurrency communities. Some projects even have their own official channels where developers and community members interact.
- **BitcoinTalk**: A long-standing forum dedicated to cryptocurrency discussions. BitcoinTalk can be a great place to analyze the general sentiment, especially when it comes to new or emerging projects.
- **Cryptopanic**: A news aggregator that also gives you a snapshot of public sentiment toward a cryptocurrency.

As you read through these conversations, focus on the general attitude of the community. Is there excitement and optimism? Or is there growing skepticism and fear? You should also engage with the community, ask questions, and gauge the tone of the responses. Be cautious, though—make sure the discussions are coming from real users and not bots or paid promoters.

Example: Analyzing Ripple (XRP) Sentiment

If you were analyzing Ripple (XRP), here's how you'd go about sentiment analysis:

1. **Go to Twitter**: Look up hashtags like #XRP or $XRP. See if the tone is predominantly positive (e.g., "XRP to the moon!") or negative (e.g., "XRP lawsuit drama ruining its future"). This can give you an immediate sense of how the market feels about the project.
2. **Check Reddit Discussions**: Visit the r/XRP or r/CryptoCurrency subreddits to see what people are saying. If the overall sentiment is highly positive, that may indicate a bullish trend. However, if many users express concerns about regulatory issues or the ongoing SEC lawsuit, that could be a bearish signal.
3. **Explore BitcoinTalk or Other Forums**: Look at long-form discussions to get a deeper insight into what experienced traders and investors think. Are they bullish on Ripple's tech and its potential future growth, or are they concerned about the ongoing legal battles?

Final Thoughts: Sentiment as a Guide, Not a Guarantee

Sentiment analysis is a fantastic way to improve your decision-making process, but it's important to remember that it's not foolproof. It's just one piece of the puzzle. Sentiment can be volatile, and sudden shifts in market mood can happen at any moment. That's why it's essential to combine sentiment analysis with other methods like fundamental analysis and technical analysis to get a full picture of the market.

In short, sentiment is a reflection of public perception. And when you can harness it, you can position yourself better to capitalize on trends, avoid mistakes, and make well-informed decisions. Whether you're looking to buy, sell, or hold, sentiment analysis can help you navigate the ever-changing cryptocurrency market with confidence.

By combining sentiment with data, you're not just relying on your gut feeling—you're using community-driven insights to guide your decisions. So, remember, keep your ears to the ground, your eyes on the charts, and your mind sharp. The crypto market waits for no one, but with sentiment analysis, you can make sure you're always in the know.

Technical Analysis in Crypto: A Guide for Short-Term Success

As you continue to refine your approach to cryptocurrency investments, there are several powerful tools that can help you navigate the ever-evolving market. After diving into Fundamental Analysis (FA) and Sentiment Analysis (SA), the next tool to add to your trading arsenal is Technical Analysis (TA). Technical analysis focuses on price movements, historical data, and various charting patterns, making it essential for those looking to capitalize on short-term opportunities, such as day trading and swing trading.

Here's how you can use Technical Analysis effectively:

1. **Use LunarCrush for Social Mentions and Data Analysis:** LunarCrush is a powerful tool that tracks social media mentions and sentiment for specific cryptocurrencies. By tracking these mentions in bulk, you can get a quick overview of how a particular coin or token is trending in social circles. This tool aggregates data from multiple platforms and provides a snapshot of how much public attention and buzz a project is generating. This information can give you a better idea of the momentum behind a coin, which is crucial for short-term trading decisions.

2. **Utilize IntoTheBlock for In-Depth Analytics:** IntoTheBlock is an artificial intelligence-driven platform that offers deep analytics about blockchain projects, including on-chain data, social metrics, and market conditions. By using this tool, you can analyze not only social sentiment but also key indicators like token distribution, volatility, and transaction patterns. This allows you to gauge the stability of a cryptocurrency before making a move, as well as identify patterns that might indicate upcoming price action.

3. **Stay Updated Through Industry News Portals:** Reading industry-related media portals like CoinDesk and CoinTelegraph can help you stay up to date with the latest market news and trends. These media outlets often provide breaking news, project developments, and expert analyses that are vital for both long-term and short-term investors. For example, news about regulatory changes or new partnerships can trigger market reactions, and being among the first to know can give you a trading advantage.

4. **Track Upcoming Events on CoinMarketCal:** CoinMarketCal is a great resource for tracking upcoming events related to specific cryptocurrencies. These events include product launches, token burns, community meetups, or protocol upgrades. Often, these events lead to significant price fluctuations, especially when the community is anticipating big news. By staying informed about these events, you can time your trades more effectively, capitalizing on price movements caused by hype or market reactions.

5. **Monitor Rumors, Regulations, and News on Cryptopanic:** Cryptopanic aggregates the latest news and rumors surrounding cryptocurrency projects. It's an excellent resource for keeping up with the latest developments in the crypto space, as news can drastically impact market sentiment. Regulations, partnerships, technological breakthroughs, or security breaches can all

affect price action, and staying updated on these events will help you make more informed trading decisions. **6. Participate in Community Discussions on Reddit**

Reddit is home to many active cryptocurrency communities where users share news, updates, and analysis of various coins. Subreddits like r/CryptoCurrency, r/Bitcoin, and r/Altcoins often contain valuable insights from experienced traders. By reading through discussions and community feedback, you can get a sense of how the market feels about certain projects, and it can also provide clues to potential market movements. Engaging with the community directly can help you filter out noise and identify which coins are gaining traction.

7. **Check the Fear & Greed Index for Market Sentiment:** The Fear & Greed Index is a sentiment tool that measures the overall mood of the cryptocurrency market. A high level of greed often signals that a coin may be overbought, while extreme fear can indicate a buying opportunity. By regularly checking this index, you can gauge market sentiment and adjust your trading strategy accordingly.

8. **Use the "Social" Tab on CoinMarketCap or CoinGecko:** Both CoinMarketCap and CoinGecko feature a "Social" tab that aggregates discussions, social media mentions, and news articles about a cryptocurrency. By exploring these platforms, you can see how people are talking about a particular project and gain insights into potential trends. This is especially useful for spotting emerging projects before they hit mainstream attention.

How Technical Analysis Complements Other Research Methods

Now that you've equipped yourself with tools for Fundamental and Sentiment Analysis, Technical

Analysis (TA) serves as the final piece of the puzzle. TA focuses primarily on price trends and volume data. It removes the "human" element from analysis, allowing traders to use charts and indicators to predict future price movements based on past performance.

While Fundamental Analysis is best suited for long-term predictions, and Sentiment Analysis works well for mid-term decisions, Technical Analysis is invaluable for identifying short-term opportunities and making timely trades. Here's why:

1. **Identifies Entry and Exit Point**s: TA helps you find the optimal points to enter and exit the market, whether you're day trading or swing trading.
2. **Supports Risk Management**: By analyzing support and resistance levels, traders can set stop-loss and take-profit orders to protect their investments and lock in profits.
3. **Detects Price Patterns and Trends**: Candlestick patterns, moving averages, and chart formations like head and shoulders or triangles can provide early signals of price direction.
4. **Boosts Confidence for Day Traders**: Day traders rely heavily on TA because it offers a more objective approach to decision-making, reducing the impact of emotions and biases.

In summary, mastering the combination of Fundamental Analysis, Sentiment Analysis, and Technical Analysis can significantly improve your ability to make well-informed trading decisions. Each analysis type offers unique insights that can guide you toward more profitable outcomes in the fast-paced crypto market.

Technical Analysis in Cryptocurrency: Tools and Benefits

The crypto markets tend to repeat historical price patterns. The saying "history tends to repeat itself" is particularly true in the cryptocurrency markets. Patterns that have occurred in the past often resurface, giving traders and investors a historical context for what might happen next. By observing the price movements of a cryptocurrency over time, you can spot recurring trends or patterns that give you valuable insight into future price behavior.

Predicting Price Movements with Charting Patterns

One of the key strengths of technical analysis is that it allows you to predict small upcoming price movements with a higher degree of accuracy. This is possible because many traders rely on similar tools and charting methods to make decisions, which in turn creates repeatable patterns. Some of the most popular indicators and patterns used for technical analysis include:

- **Moving Averages (MA) and Exponential Moving Averages (EMA)**: These help smooth out price data to identify the direction of the trend.
- **Fibonacci Retracements:** These levels help identify potential points where a market might reverse after a strong trend.
- **Smart One Index (SOI)**: This index tracks momentum to identify trend strength.
- **Bollinger Bands**: Used to measure volatility and determine whether an asset is overbought or oversold.
- **Relative Strength Index (RSI)**: Indicates whether an asset is overbought or oversold by comparing the magnitude of recent gains to recent losses.
- **Elliot Waves**: A form of wave analysis used to predict the cyclical nature of markets.

- **Harmonic Patterns**: These identify potential reversal points in the market by using geometrical patterns and ratios.

By studying these indicators, you can anticipate whether the market is about to enter a bullish or bearish trend and make informed decisions about your trades. These indicators often serve as a "map" of future price action.

Identifying Oversold or Undersold Conditions

Another advantage of technical analysis is its ability to spot when a cryptocurrency is oversold or undersold. This is especially helpful for traders who want to buy low or sell high. Key indicators such as RSI, Bollinger Bands, and Stochastic Oscillators can show when an asset is either oversold (a long wick on a candlestick pattern or RSI below 30) or oversold (RSI above 70), which signals potential price reversals.

Traders can use this information to make smarter decisions about when to exit a position or when to enter a position, potentially locking in profits at the peak and avoiding losses during downtrends.

Technical Analysis as a Projection Tool

Technical analysis also serves as a projection method that uses a variety of indicators and chart patterns to analyze market trends and predict where prices might go in the short term. This makes it an essential tool for traders looking to capitalize on price movements in the cryptocurrency market, especially when the market is volatile.

Best Platforms for Cryptocurrency Technical Analysis

There are several platforms available where you can conduct in-depth technical analysis for cryptocurrency trading. Two of my personal favorites are SmartTrader and TradingView, both of which offer great functionality and insights for both beginners and seasoned traders.

1. **SmartTrader**: SmartTrader is a user-friendly platform that offers advanced charting tools and a wide range of technical indicators to help traders make informed decisions. It allows you to customize your charts, set up alerts, and analyze market trends efficiently.

2. **TradingView**: TradingView is one of the most popular charting platforms in the world. It offers a massive range of indicators, drawing tools, and chart types. You can track specific cryptocurrencies and apply technical analysis indicators such as moving averages, Bollinger Bands, Fibonacci retracement levels, and much more. TradingView also allows you to paper trade (practice trading without real funds) before you take the plunge with your own capital. Once you're confident in your skills, you can link your exchange accounts to TradingView and begin trading live.

Both platforms also offer the option to access premium features, but even the free versions of these platforms provide solid functionality for traders looking to improve their analysis skills.

Other Useful Platforms for Technical Analysis

1. **Binance**: Binance offers its own advanced charting and trading tools, including a wide range of indicators that are essential for conducting detailed technical analysis.

2. **Coinigy**: Coinigy is another platform offering a variety of charting tools and analytics, along with integration with over 40 exchanges for real-time data and trading.

3. **Cryptowatch:** Owned by Kraken, Cryptowatch provides an easy-to-use charting platform with a focus on real-time data from multiple exchanges. It's great for traders looking to track cryptocurrency prices and execute trades with ease.

4. **Messari:** Messari provides both fundamental and technical analysis tools. It's an excellent platform for those looking to get deep insights into market trends, tokenomics, and charts.
5. **CoinMarketCap & CoinGecko**: While these platforms are mostly known for tracking cryptocurrency prices and market data, they also offer a variety of charting tools, along with community data, social sentiment, and more.
6. **TradingView Alternatives**: You can also consider platforms like ChartStar or Krypto if you're looking for more specific tools designed for cryptocurrency charting and technical analysis.

By combining technical analysis, fundamental analysis, and sentiment analysis, you are setting yourself up to make well-informed decisions in the crypto markets. Each type of analysis has its own strengths, and by understanding how to use each method effectively, you can build a more complete trading strategy that minimizes risk while maximizing profit opportunities.

The cryptocurrency market is fast-paced and volatile, and the ability to assess and react to market conditions quickly is vital. Technical analysis, with its ability to predict short-term movements and identify price trends, is especially crucial for day traders and swing traders who need to make quick decisions.

As with any trading strategy, it's important to practice and refine your skills over time. Start with paper trading, and when you feel confident, you can move on to live trading with real funds. Above all, always remember that no analysis method is foolproof—having a sound risk management strategy in place will help ensure that you stay on track and achieve long-term success.

Crypto Trading Tools and Platforms for Technical Analysis

When it comes to executing your cryptocurrency trading strategies, having access to reliable tools for technical analysis is essential. Fortunately, there are several platforms available that can help you track trends, spot price movements, and improve your trading decisions. Here are a few top platforms and tools to aid you in your technical analysis journey:

- **CryptoWatch**: CryptoWatch, a platform recently acquired by US-based exchange Kraken, is a powerful tool designed for digital asset-focused chart analysis. It offers a comprehensive range of technical analysis tools that can help traders make informed decisions. With the CryptoWatch terminal, you can easily track your portfolio, analyze charts, and even execute trades directly on the Kraken exchange. This integration makes it a seamless option for both beginners and experienced traders looking to streamline their trading experience.
- **Coinigy**: Coinigy is another excellent platform that positions itself as "the ultimate cryptocurrency portfolio management suite for professional traders." It has earned its reputation for being an all-in-one tool for crypto traders and investors. One of the standout features of Coinigy is its ability to connect with nearly every crypto exchange, allowing users to manage multiple accounts from a single dashboard. In addition to portfolio tracking and management functions, Coinigy provides advanced charting tools, execution capabilities across various exchanges, and wallet management. This platform is a must-try for anyone serious about cryptocurrency trading.

Crypto Trading Bots

For those looking to automate their trading strategies, crypto trading bots are a great solution. Trading bots can execute your planned trades automatically, allowing you to capitalize on market movements while you focus on other activities. Bots can remove emotions from trading, reduce the risk of manual errors, and ensure that your strategy is followed consistently.

Two of the most popular free, open-source trading bots are Gekko and ZenBot, both of which offer a range of useful features for crypto traders:

- Gekko (launched in 2013) is one of the first and still one of the most popular cryptocurrency trading bots available. It allows users to build, backtest, and execute cryptocurrency trading strategies based on various technical indicators. Gekko is highly customizable and has a wide range of supported exchanges, making it a great choice for developers and traders alike.
- ZenBot (first launched in 2016) has evolved significantly over the years. With a growing community of developers, ZenBot offers backtesting, live execution of custom trading strategies, and a paper trading mode. Paper trading allows users to simulate trading without risking real funds, making it a fantastic option for those new to crypto or testing new strategies before committing real capital.

Top Resources for Cryptocurrency Market Research

As we've discussed throughout this guide, thorough research is essential to navigating the cryptocurrency market successfully. To help you on your journey, I've compiled a list of some of the top websites and platforms where you can conduct deep, thorough market research across the three critical areas: Fundamentals, Sentiment, and Technical Analysis. These tools will give you the insights and data necessary to make smarter investment decisions and minimize risk.

Here are some of the best sites you should start utilizing:

Fundamental Analysis

- Coingecko https://www.coingecko.com/
- Coinmarketcap https://coinmarketcap.com/
- Coincheckup https://coincheckup.com/
- Coinpaprika https://coinpaprika.com/
- Lunarcrush https://lunarcrush.com/markets
- Intotheblock https://www.intotheblock.com/
- Fiatleak https://fiatleak.com/
- Smarttrader https://smarttrader.com/
- Coinmarketcal https://coinmarketcal.com/en/

Sentiment Analysis

- Cryptopanic https://cryptopanic.com/
- Coindesk https://www.coindesk.com/
- Cointelegraph https://cointelegraph.com/
- Crypto Fear & Greed Index https://alternative.me/crypto/fear-and-greed-index/
- TodayOnChain https://todayonchain.com/
- News BTC https://www.newsbtc.com/
- Bitcoin Magazine https://bitcoinmagazine.com/
- CryptoSlate https://cryptoslate.com/

- Bitcoinist https://bitcoinist.com/
- Forbes Crypto https://www.forbes.com/crypto-blockchain/
- Null Tx https://nulltx.com/
- Bitcoin.com https://www.bitcoin.com/
- Invest In Blockchain https://www.investinblockchain.com/
- Blockonomi https://blockonomi.com/
- Coinspeaker https://www.coinspeaker.com/

Technical Analysis

- Coinigy https://www.coinigy.com/
- Nomics https://nomics.com/
- Cryptocompare https://www.cryptocompare.com/
- Bitinfocharts https://bitinfocharts.com/
- Prices.org https://prices.org/
- CryptoWatch https://www.cryptowat.ch/
- TradingView https://www.tradingview.com/
- RSI Hunter https://rsihunter.com/

Why These Sites Matter

These resources will provide you with the tools needed to understand a cryptocurrency from all angles:

- **Fundamental Analysis**: Learn about the team, project details, partnerships, market cap, circulating supply, and more to ensure you're investing in something valuable.
- **Sentiment Analysis**: Gauge public opinion, analyze hype, and understand market trends and sentiment. By staying ahead of the curve, you can make timely decisions that capitalize on market sentiment.
- **Technical Analysis**: Dive into price charts, volume trends, indicators, and candlestick patterns to predict price movements and find ideal entry and exit points for short-term trades.

A Comprehensive Research Approach

Now that you have access to a list of resources, the real challenge is to use them wisely. The most successful investors spend hours researching every aspect of a cryptocurrency before they even consider investing.

Here's how I approach my research process:

1. **Community Support**: I first check if the community around a project is active and passionate. A dedicated community can give projects the staying power they need to survive market volatility.
2. **Utility and Use Case**: I always ask: *Does this project have a real-world utility?* The more practical applications a cryptocurrency has, the better its chances for long-term success.
3. **Social Media and Whitepapers:** I review their social media channels, website, whitepaper, and roadmap to ensure the team has a solid plan for the future. If they're not transparent or lacking in communication, it's a red flag.

4. **Charts and Historical Data**: I analyze past performance, support and resistance levels, and key technical indicators to get a clearer picture of the cryptocurrency's potential.

By performing this multi-pronged approach, you minimize the risk of falling for scams like rug-pulls, honeypots, or projects that have no real future. It also helps you to find hidden gems with strong potential early on, before the masses dive in.

Remember, market research isn't just about gathering information – it's about using that information to make informed, logical decisions. The more time and effort you dedicate to researching and learning, the better your chances of navigating the crypto space successfully.

So, how badly do you want financial freedom? Is it more than you want to breathe? If so, start your research journey today, and keep learning every day. The market is full of opportunities, and with proper research, you'll be better equipped to spot them before everyone else.

Good luck on your journey and may your research lead you to success!

Chapter 23

Affiliating Marketing: The Power of Referring a Friend

Have you ever been to a restaurant so amazing that you couldn't stop talking about it to everyone around you? The food was delicious, the atmosphere was peaceful, and the vibe was just perfect. Every time you went, you felt like you were part of something special. And then, you find out that the restaurant offers you a 15% discount for referring them to friends. Of course, you're going to refer all your friends to that restaurant, because why wouldn't you want them to experience the same amazing food and atmosphere?

That's affiliate marketing at its core—companies or businesses rewarding you for bringing new customers their way. It's something that works across every industry, and cryptocurrency is no exception. What if I told you that you can earn cryptocurrency by simply referring your friends and family to certain exchanges, platforms, or even services like hardware wallets?

In this chapter, we will explore the best crypto websites offering the most lucrative referral programs where you can earn passive income by introducing your loved ones to the world of crypto. Let's dive in!

Before You Start: Setting Up for Success

Before you can start earning, you need to create an account on the platforms offering referral programs. Setting up your account is simple—just enter your name, email, and verify your email address. If you don't complete these steps, you won't be able to earn from your referrals. Once you're signed up, you'll be ready to start sharing your referral links and making some crypto income!

Top Crypto Platforms for Referral Programs

Here are some of the best cryptocurrency platforms with referral programs that can help you earn by sharing with your friends and family:

1. **Binance:** Binance is one of the largest and most reputable cryptocurrency exchanges globally, and they offer a robust referral program. Here's how it works:

   - How to Get Started: Register on Binance and generate your unique referral link or scannable QR code. To do this, go to "Account" and click on "My referral ID." On the referral page, you'll see a list of how many friends you've referred and how much you've earned.
   - Payouts: Binance refreshes your rewards every 24 hours at 12:00 AM UTC.
   - What You Earn: You earn between 20%-40% of your friend's trading fees.
   - What Your Friends Need to Do: They simply need to sign up and start trading to generate rewards for you.
   - What Your Friends Get: Your friends can get rewards, too, depending on how much you decide to share. If your wallet balance is less than 500 BNB, you can choose to share 5%-10% of your commission. For those with more than 500 BNB in their balance, you can share 5%-20% of your commission.

2. **Coinbase:** Coinbase is one of the most popular cryptocurrency exchanges in the United States, and their affiliate program is simple to use. Here's how you can get started:

   - How to Get Started: On the web version, head over to the "Referrals" page. If you're using the app, tap on "Settings," then find the referral link or code and share it.
   - Payouts: Coinbase doesn't have a fixed bulk trading requirement. As long as your friends sign up and trade within six months, you'll earn your commission.

What You Earn: Coinbase rewards you with a percentage of your friend's trades. While it doesn't specify exact amounts, your earnings increase based on your referral activity and your friends' trading volume.

- What Your Friends Need to Do: Your friends need to sign up and trade, and you'll earn a commission based on their activity.
- What Your Friends Get: Your friends will receive a bonus or discount for signing up through your referral link.

3. **KuCoin:** KuCoin is another top exchange that offers a fantastic referral program, which is great for those who want to promote crypto trading in their community. Here's how it works:

- How to Get Started: Once you've signed up for KuCoin, navigate to the "Referral Program" section in your account settings to get your unique referral link.
- Payouts: KuCoin pays you a percentage of your friend's trading fees, typically around 20%-40%.
- What You Earn: You'll earn a commission based on your referrals' activity, and you can increase your rewards by sharing with more people.
- What Your Friends Need to Do: Your friends need to sign up and start trading to generate rewards for you.
- What Your Friends Get: Your friends receive a reward for signing up through your referral link, usually in the form of trading fee discounts or a bonus.

4. **Gemini:** Gemini is a regulated exchange based in the United States, and they offer a lucrative referral program that pays you for referring new users. Here's how you can participate:

How to Get Started: After signing up for Gemini, go to the "Earn Rewards" section in the account settings, and you'll receive a referral link to share.

- Payouts: Gemini offers a flat fee per referred user, usually based on how much the referred user deposits into their account.
- What You Earn: You earn a commission based on the deposits made by your referred friends. It's a simple process of earning a flat fee for each new account you bring in.
- What Your Friends Need to Do: Your friends must sign up using your referral link and make a deposit into their Gemini account.
- What Your Friends Get: New users often receive a sign-up bonus or trading fee discounts as an incentive to join through your referral.

5. **Ledger (Hardware Wallet):** Ledger is one of the leading manufacturers of hardware wallets, offering a secure way for users to store their cryptocurrency. They have an affiliate program where you can earn commissions by referring people to purchase their products.

- How to Get Started: Sign up for the Ledger Affiliate Program, and you'll get access to banners, links, and referral codes that you can share.
- Payouts: You earn a percentage of the sales made through your referral link. This typically ranges from 10%-30% based on the product.
- What You Earn: You earn a commission for each sale made through your referral link. The more people you refer, the higher your earnings.
- What Your Friends Need to Do: Your friends need to purchase a Ledger hardware wallet using your

What Your Friends Get: Some promotions or discounts may be available depending on when they make their purchase.

6. **Crypto.com:** Crypto.com is another giant in the crypto world with over 25 million users. They offer an attractive referral program, particularly for those who want to stake tokens and earn rewards.

   - Payout: Immediate.
   - What you get: Up to $10-$2000 in CRO (Crypto.com's native token), depending on how much CRO your friend stakes. You also receive up to 50% of your friend's net trading fees.
   - What your friends need to do: Complete the KYC process, stake at least 1000 CRO, and trade on the platform.
   - What your friends get: No rewards for them in this program, but they'll get access to the platform's features.

7. **ZenGo:** ZenGo is a crypto exchange with a focus on ease of use. The platform is available in 100+ countries and has a straightforward referral system that allows you to earn Bitcoin.

   - Payout: Weekly, in Bitcoin.
   - What you get: 50% of the commission on every crypto purchase your referred users make.
   - What your friends need to do: Download the app, sign up, and make a purchase.
   - What your friends get: $10 worth of Bitcoin when they make their first crypto purchase on ZenGo.

     Additionally, purchases over $200 in crypto earn them a BTC cashback.

8. **Paxful:** Paxful is a peer-to-peer (P2P) platform that lets users buy and sell Bitcoin directly with one another. Paxful also has an excellent affiliate program, rewarding you for both direct and indirect referrals.

   Payout: Instant once you reach $10 in earnings.

   - What you get: 50% of the escrow fees from your direct referrals, plus 10% of escrow fees from anyone they refer.
   - What your friends need to do: Sign up and start trading Bitcoin.
   - What your friends get: They will receive the same benefits that you enjoy on your own referrals.

9. **LocalBitcoins:** LocalBitcoins is a peer-to-peer marketplace that connects buyers and sellers of cryptocurrency directly. It's widely used for buying and selling Bitcoin without a middleman. Through the LocalBitcoins referral program, you can earn 20% of the commission fee for every transaction completed by a user who signed up through your referral link.

   - Payout: 20% of the commission fee for each transaction made by the user referred by you.
   - What You Get: 20% of the transaction commission.
   - What Your Friends Need to Do: Download the app, sign up, and start buying or selling cryptocurrency.
   - What Your Friends Get: They will receive the same 20% commission share on transactions made by users they refer.

10. **Changelly:** Changelly is an instant crypto-to-crypto exchange platform that allows users to swap different cryptocurrencies quickly. Their affiliate program offers a generous 50% lifetime commission for every transaction made by users referred through your link. Traders and investors can also embed Changelly's widget on their website to increase their earnings.

- Payout: 50% lifetime commission for each transaction made by a referred user.
- What You Get: 50% of the revenue generated from the transactions made by your referrals. What Your Friends Need to Do: Sign up, complete the KYC process, and exchange cryptocurrencies on Changelly.
- What Your Friends Get: They get the same benefits, with a share of the commission for their own referrals.

11. **Hashflare:** Hashflare is a cloud mining service that allows users to mine cryptocurrencies by purchasing mining contracts. The platform offers a 10% commission for affiliates who refer new customers that purchase mining contracts. This program allows you to earn passive income through mining, without the need for physical hardware.

- Payout: 10% commission for users who purchase mining contracts through your referral link.
- What You Get: A 10% commission for the mining contracts purchased by your referrals.
- What Your Friends Need to Do: Sign up, purchase a mining contract, and start mining. What Your Friends Get: Same 10% commission on any subsequent referrals they make.

12. **Xcoins.com:** Xcoins.com is a marketplace where users can buy cryptocurrencies. Their affiliate program offers generous rewards to those who refer others to the platform. For every user that signs up using your link, they'll get $10, and you will also earn $10. Plus, you'll receive a 20% lifetime commission from the fees paid by anyone you refer.

- Payout: $10 for each user you refer, plus 20% of the fees paid by your referrals
- What You Get: $10 for each user, along with 20% of the fees paid by the referred user
- What Your Friends Need to Do: Download, sign up, and buy crypto What Your Friends Get: $10 for signing up, along with the same percentage rewards you enjoy from their referrals

13. **Quadency:** Quadency is a multi-exchange crypto automation trading platform, ideal for users who want to automate their trading strategies. Their referral program allows you to earn up to $200 per successful referral, and you'll also be rewarded with a free PRO plan upgrade as you refer more users. Rewards are paid out in Bitcoin, which is a nice touch for crypto enthusiasts.

- Payout: Up to $200 for each successful referral
- What You Get: Up to $200 per successful referral and free PRO plan upgrades
- What Your Friends Need to Do: Download, sign up, and start trading
- What Your Friends Get: The same percentage rewards as you for referring others

14. **Ellipal:** Ellipal is another top hardware wallet provider known for its advanced security features. They offer a competitive affiliate program that allows you to earn between 10%-15% commission on each successful referral.

- Payout: 10%-15% of the product cost as a referral commission
- What You Get: 10%-15% commission for each sale made via your referral link
- What Your Friends Need to Do: Sign up, purchase a hardware wallet
- What Your Friends Get: Same referral commission percentages you enjoy

15. **Covesting:** Covesting is a fully licensed cryptocurrency exchange and trading platform that also supports copy trading. If you're a trader or investor who likes to share your strategies with others,

Covesting's affiliate program could be a great way to earn passive income. They offer up to 40% commission on your referrals' trading activity.

- Payout: 0% to 40% of trading commissions, based on performance
- What You Get: Real-time commission payments, ranging from 20%-40%
- What Your Friends Need to Do: Sign up, engage in crypto trading or copy trading
- What Your Friends Get: Same referral rewards you earn for referring others

16. **Bitpanda:** Bitpanda is a platform that enables users to buy and sell Bitcoin and other cryptocurrencies using methods like credit cards, PayPal, and wire transfers. They offer a lifetime revenue share of 20% for every transaction made by a user referred through your link.

- Payout: 20% of each transaction made by a referred user
- What You Get: 20% revenue share from the transactions made by your referrals
- What Your Friends Need to Do: Sign up and make a crypto purchase
- What Your Friends Get: The same rewards you enjoy from referring others

17. **Trezor:** Trezor, like Ledger, is another reputable name in the world of cryptocurrency hardware wallets. They offer a great affiliate program with commissions ranging from 12%-15% for every product sold via your referral link.

- Payout: 12%-15% commission on product cost for each sale
- What You Get: 12%-15% commission on every Trezor wallet sold through your referral link
- What Your Friends Need to Do: Sign up and purchase a hardware wallet
- What Your Friends Get: Same referral rewards as you

18. **KeepKey:** KeepKey is another popular cryptocurrency wallet known for its user-friendly features and secure storage options. KeepKey offers a lucrative referral commission, providing affiliates with 25% of the product price for each sale made through their referral link.

- Payout: 25% value of the product price as a referral commission.
- What you get: 25% value of the product price as a referral commission.
- What your friends need to do: Download, sign up, and buy crypto.
- What your friends get: Same percentages you enjoy on their referrals.

19. **Indacoin:** Indacoin is an instant crypto exchange that allows users to purchase over 100 cryptocurrencies using a credit card. They offer a solid referral program, providing affiliates with 3% of the amount invested by the referred user.

- Payout: 3% of the amount user invested through your referral link.
- What you get: 3% of the amount user invested through your referral link.
- What your friends need to do: Download, sign up, and buy crypto.
- What your friends get: Same percentages you enjoy on their referrals.

20. **CEX.io:** CEX.io is a well-established cryptocurrency exchange platform that offers a variety of services, including brokerage services and margin trading. They provide a lifetime revenue share for every transaction made by a user referred through your link, ensuring continuous earnings.

- Payout: Lifetime revenue share from the commission earned on each transaction made by a user referred through your link.

- What you get: Lifetime revenue share from the commission earned on each transaction made by a user referred through your link.
- What your friends need to do: Download, sign up, and buy crypto.
- What your friends get: Same percentages you enjoy on their referrals.

21. Wirex

Wirex is leading crypto debit card provider which allows users to make purchase for their Bitcoin, Ethereum, Litecoin etc through debit card. Depending upon the traffic quality and Geo location they offer up to $10 for each sale made through your referral link.

- Payout: $10 for each sale made through your referral link.
- What you get: $10 for each sale made through your referral link.
- What your friends need to do: Download, sign up and buy crypto.
- What your friends get: Same percentages you enjoy on their referrals.

22. CoinMama: Coinmama is an easy-to-use platform where users can instantly buy cryptocurrencies with a credit card. Their referral program is quite attractive, sharing 15% of the commission fee on every purchase made through your referral link.

- Payout: 15% of the commission fee on every purchase made through your referral link.
- What you get: 15% of the commission fee for each purchase made by users you refer.
- What your friends need to do: Download, sign up, and buy crypto.
- What your friends get: The same referral benefits you enjoy when they make purchases.

23. **PrimeXBT:** PrimeXBT is a renowned crypto margin trading platform that offers industry-leading leverage of up to 100x. They provide a lucrative referral program where you earn 20% of the commission fees collected per transaction.

- Payout: 20% of the commission fee collected per transaction.
- What you get: 20% of the commission fee on every transaction made by a user you refer.
- What your friends need to do: Download, sign up, and make a trade.
- What your friends get: The same commission share benefits as you when they engage in trades.

24. **IQ Mining:** IQ Mining is a cloud mining platform that allows users to mine Bitcoin and other altcoins using cloud-based infrastructure. As part of their affiliate program, they offer up to 18% of the revenue share through their partner affiliate network, cryptoffers.com.

- Payout: 18% of the revenue share via the partner affiliate network.
- What you get: 18% of the revenue share for each referral through the partner network.
- What your friends need to do: Download, sign up, and start mining.

25. **Poloniex:** Poloniex is a popular cryptocurrency exchange known for offering a wide variety of crypto pairs and trading options. Through its referral program, Poloniex rewards both you and your referrals for bringing in new users. You will earn 20% of your referral's trading fees, while your referral will receive a 10% discount on trading fees by using your referral link. The rewards are paid out daily in USDC and can be earned for up to 180 days after your referral signs up, or until both of you collectively reach $5,000 in fees.

- Payout: 20% off of your referral's trading fees while they also earn a 10% discount for using your referral link.

- What you get: 20% of your referral's trading fees, paid daily in USDC, while your referral enjoys 10% discount on their fees.
- What your friends need to do: Download, sign up, and start trading on Poloniex.
- What your friends get: A 10% discount on their trading fees by using your referral link. They also become eligible to refer others and earn their own rewards.

26. **BitMEX:** BitMEX is a leading cryptocurrency derivatives exchange that specializes in margin trading with high leverage. BitMEX offers an attractive affiliate program with a commission structure that rewards you based on the total trading volume generated by your referrals. The more your referrals trade, the higher the commission rate you can earn, ranging from 10% to 20%.
    - Payout: 10% to 20% commission depending on the total turnover generated by your referrals.
    - What you get: Between 10% and 20% of the trading fees generated by your referrals, based on the total turnover.
    - What your friends need to do: Download, sign up, and start trading on BitMEX.
    - What your friends get: Same referral benefits you enjoy, enabling them to earn commissions by referring others as well.

27. **CoinMetro:** CoinMetro is a user-friendly cryptocurrency exchange that offers a robust affiliate program, rewarding you for introducing friends and family to their platform. Through CoinMetro's program, you earn a commission from your direct referrals and even from the people they refer. You will earn a generous 40% commission from your direct invitees' trading fees, and 10% from the referrals of your referrals. It's a simple way to earn a continuous passive income by sharing the benefits of trading and investing in crypto.

- Payout: 40% commission from your direct invitees, plus 10% from the people your referrals bring in.
- What you get: 40% of the trading fees from your direct referrals and 10% of the trading fees from the people your referrals bring in.
- What your friends need to do: Download, sign up, and start trading on CoinMetro.
- What your friends get: The same referral benefits you enjoy, allowing them to earn by referring others as well.

28. BitFlyer: BitFlyer, one of the largest cryptocurrency exchanges based in Japan, offers an easy-to-use referral program. When you refer someone, both you and the person you refer will receive roughly $12 (10 euros) in Bitcoin after they create an account and complete a minimum trade of $120 (100 euros). This is a great way to introduce others to the world of cryptocurrency while earning some Bitcoin as a reward for your efforts.

- Payout: Referrers and referrals receive $12 (10 euros) in Bitcoin once the referred user makes a trade of at least $120 (100 euros).
- What you get: $12 (10 euros) in Bitcoin for every successful referral who trades at least $120 (100 euros).
- What your friends need to do: Download, sign up, and make a trade of $120 (100 euros) or more.
- What your friends get: $12 (10 euros) in Bitcoin for completing their first trade.

Tips for Success:

1. **Promote on Multiple Platforms**: Share your referral link on social media, blogs, or crypto-related forums to reach a wider audience.

2. **Educate Your Referrals**: Help your friends understand how to use the platform so they feel more comfortable signing up and using the services.
3. **Stay Updated on Program Changes**: Crypto affiliate programs can change over time, so it's important to stay updated with the latest offers and terms.

Remember to always do your due diligence to ensure the platforms you're referring people to are credible and trustworthy. This will help you build a solid reputation, and over time, you can create a steady stream of passive income.

Affiliate marketing in crypto is a fantastic way to get involved in the ecosystem, earn cryptocurrency, and introduce your friends and family to the world of digital finance. So, which platform will you start with? The crypto world is full of opportunities, and now you know how to make the most of them!

Happy referring, and may your crypto journey be rewarding!

Chapter 24

The Biggest Transfer of Wealth in the Age of Cryptocurrency Outro

We are living in an era where technology has revolutionized the way we live, work, and interact with the world. With mobile phones in our pockets, access to the internet, and platforms that connect us to global opportunities, it's never been easier to be successful. In fact, young people under 16 years old are already becoming millionaires thanks to platforms like YouTube, Twitch, TikTok, and Instagram. Add to that the explosive growth in cryptocurrency, blockchain technology, and NFTs, and we can safely say that the age of financial freedom and digital innovation is upon us.

The Possibilities are Limitless

With the power of a smartphone and internet access, anyone can learn how to build a business, earn passive income, or even break into the world of crypto without having a traditional degree or formal training. There's no longer a need to go to college to learn how to create something valuable in the digital economy. The information, the experts, and the resources are all available at your fingertips.

We are witnessing the largest transfer of wealth in human history. In the last year, the United States printed more money than it ever has before, leading to inflation and a devaluation of the US Dollar. This is not a small shift, but a profound one that will have lasting effects on the global economy. The US Dollar's dominance in global markets is already weakening, with countries like Saudi Arabia, South Korea, Russia, Venezuela, and India exploring alternatives to the Dollar, such as the

Chinese Yuan and Digital Currencies

China, for instance, is already leading the way with its Digital Yuan. In fact, China's Digital Yuan has already processed over $87.2 billion in transactions and is making strides to replace traditional cash with a fully digital economy. The Beijing Winter Olympics saw the successful rollout of this currency, further solidifying China's ambitions to dominate the global financial landscape.

The shift towards digital currencies is happening. In addition to the Digital Yuan, other nations like El Salvador have adopted Bitcoin as legal tender, with more countries including Brazil, Venezuela, and Argentina moving towards cryptocurrency legalization. In fact, even countries facing sanctions, like Russia, are increasingly turning to cryptocurrency to bypass traditional banking systems, as evidenced by the $215 billion estimated in cryptocurrency holdings in Russia.

But what does this mean for you? How can you position yourself for success in the coming era of cryptocurrency and blockchain technology?

The Risk of Being Left Behind

As cryptocurrency and digital currencies continue to grow, those who resist the change or fail to adapt risk being left behind. An experimental study conducted in El Salvador showed that many of the country's poorest citizens, especially older generations and those in rural areas, are finding it difficult to transition to a digital currency system. With limited access to smartphones, the internet, and ATMs, they may struggle to keep up with a rapidly changing financial landscape.

As technology and digital currencies continue to replace traditional systems, those who aren't actively learning, adapting, and participating will find themselves excluded from the opportunities available. Think about your loved ones, especially older family members who may not even know how to use a smartphone. What will they do when traditional currencies lose value or become obsolete?

This is the reason why I wrote this book — to help you, your family, and your friends embrace the new world of cryptocurrency. I want to ensure that you are prepared for the largest financial transition we've ever seen, so you don't get left behind when the future of money shifts.

The Path Forward: Learn, Adapt, and Take Action

The biggest transfer of wealth in history is unfolding, and cryptocurrency is at the forefront of that revolution. This is your chance to be part of that movement, to secure your financial freedom, and to protect yourself and your family from the financial crises that are inevitable in the traditional fiat system.

In the future, centralized banks will lose their dominance. You'll no longer be tied to a single institution to manage your money. Instead, you'll have the power to be your own bank through decentralized finance (DeFi) systems. With the right tools, you can keep your money safe, earn interest, and transact globally without relying on third-party banks.

You can also take advantage of the benefits of crypto debit and credit cards. These cards offer not just spending capabilities but rewards, travel benefits, and even insurance services that traditional banking cards don't offer. These are all opportunities for you to unlock financial freedom and use technology to improve your quality of life.

But simply reading this book and learning about cryptocurrency isn't enough. You must commit To the journey. You must take massive action to educate yourself and start using the tools and resources available. The digital revolution is moving fast, and you cannot afford to wait until it's too late.

The Call to Action

I have poured everything into this book to help you understand the opportunities in cryptocurrency and blockchain technology. My mission is to ensure that you, your family, and your friends are fully equipped to innovate and adapt in this new financial world. The world is changing, and we have to change with it.

But don't stop here. Remain a student of the game. Keep learning, keep growing, and keep expanding your network. Surround yourself with like-minded individuals who are also striving for success in the crypto space. Power in numbers is one of the most effective ways to accelerate your learning and wealth-building journey.

In the coming years, we will see the mainstream adoption of digital currencies. The question is: Will you be part of that shift? Or will you let this opportunity slip through your fingers?

As Anthony Noto, CEO of SoFi, said, "If you don't innovate, and you don't use cryptocurrency as a technology platform, you'll get left behind. Your business will be smaller. You'll be less competitive."

I am committed to helping you innovate, to helping you thrive in the digital age. Don't let this moment pass you by. Embrace the technology, seize the opportunity, and take control of your financial future.

Final Thoughts and Prayers

I pray that each person who reads this book, in the name of Jesus Christ, experiences good health, prosperity, and financial freedom. I hope that the knowledge you gain here will empower you to make wise decisions, help your loved ones, and become a beacon of success in this new digital world.

Remember to commit, take massive action, and stay humble. The road to success may not always be easy, but with the right mindset, it's more than achievable.

This is the age of technology. This is the age of cryptocurrency. The power to change your life is right in front of you.

Take action now. Your future is waiting.

The wealthiest people in the world are the biggest givers and with that I say to all happy finding and welcome to the biggest transfer of wealth in the age of cryptocurrency. I will see you guys in the metaverse.

Cheers!

Thank You

First and foremost, I want to thank you for taking the time to read this book. I truly appreciate your commitment to learning and growing in the world of cryptocurrency. Remember, the journey doesn't end here — come back to this book whenever you need guidance as you navigate the crypto markets. The more you learn, the more you grow.

As you progress, I encourage you to share this knowledge with your loved ones and friends. We can only move forward together, and helping others is the best way to reinforce your own growth.

A Few Final Words of Wisdom

- Passion + Hard Work = Success. Nothing great comes easily, but with persistence, dedication, and focus, you can achieve more than you ever thought possible.
- DYOR (Do Your Own Research) and always be mindful of your investments. Success is built on knowledge and understanding.
- Plug into a Mentor and Grow. A mentor can help guide you through the challenges and complexities of the crypto space. Learning from others' experiences can fast-track your success.
- Find a Community to Grow With. Surrounding yourself with like-minded people who are also passionate about crypto can inspire you, hold you accountable, and elevate your game.

- Take What You've Learned and Help Your Family Towards the Freedom They Deserve. One of the most powerful ways to achieve success is to bring others along for the ride. The knowledge and freedom you gain should extend to those you care about.

Stay Connected

I would love to hear about your journey and your success stories. Please feel free to reach out to me and share how this book has helped you. I'm excited to be a part of your journey to financial freedom. Email me your success stories at: apontecryptocompliance@gmail.com. I look forward to hearing from you. Wishing You the Best on Your Journey,

Xavier Aponte

www.ingramcontent.com/pod-product-compliance
Lightning Source LLC
LaVergne TN
LVHW081948200926
848484LV00009B/2528